Urban and Landscape Perspectives

Volume 19

Aims and Scope

Urban and Landscape Perspectives is a series which aims at nurturing theoretic reflection on the city and the territory and working out and applying methods and techniques for improving our physical and social landscapes.

The main issue in the series is developed around the projectual dimension, with the objective of visualising both the city and the territory from a particular viewpoint, which singles out the territorial dimension as the city's space of communication and negotiation.

The series will face emerging problems that characterise the dynamics of city development, like the new, fresh relations between urban societies and physical space, the right to the city, urban equity, the project for the physical city as a means to reveal civitas, signs of new social cohesiveness, the sense of contemporary public space and the sustainability of urban development.

Concerned with advancing theories on the city, the series resolves to welcome articles that feature a pluralism of disciplinary contributions studying formal and informal practices on the project for the city and seeking conceptual and operative categories capable of understanding and facing the problems inherent in the profound transformations of contemporary urban landscapes.

More information about this series at http://www.springer.com/series/7906

Roberta Ingaramo • Angioletta Voghera
Editors

Topics and Methods for Urban and Landscape Design

From the river to the project

Springer

Editors
Roberta Ingaramo
Department of Architecture and Design
Politecnico di Torino
Turin, Italy

Angioletta Voghera
Interuniversity Department of Regional and Urban Studies and Planning
Politecnico di Torino
Turin, Italy

Urban and Landscape Perspectives
ISBN 978-3-319-51534-2 ISBN 978-3-319-51535-9 (eBook)
DOI 10.1007/978-3-319-51535-9

Library of Congress Control Number: 2016961513

Printed on acid-free paper

This Springer imprint is published by Springer Nature
The registered company is Springer International Publishing AG
The registered company address is: Gewerbestrasse 11, 6330 Cham, Switzerland

Contents

Contributors

Alessandro Armando Department of Architecture and Design, Politecnico di Torino, Turin, Italy

Gilda Berruti Department of Architecture, Federico II University of Naples, Napoli, Italy

Grazia Brunetta Interuniversity Department of Regional and Urban Studies and Planning, Politecnico di Torino, Turin, Italy

Antonio De Rossi Department of Architecture and Design, Politecnico di Torino, Turin, Italy

Stephen K. Diko School of Planning, University of Cincinnati, Cincinnati, OH, USA

Francesco Fiermonte Interuniversity Department of Regional and Urban Studies and Planning, Politecnico di Torino, Turin, Italy

Mattia Giusiano Department of Architecture and Design, Politecnico di Torino, Turin, Italy

Roberta Ingaramo Department of Architecture and Design, Politecnico di Torino, Turin, Italy

Carlo Alessandro Manzo Department of Architecture and Industrial Design, Second University of Naples, Napoli, Italy

Paolo Mellano Department of Architecture and Design, Politecnico di Torino, Turin, Italy

Francesco Domenico Moccia Department of Architecture, Federico II University of Naples, Napoli, Italy

Guglielmina Mutani Department of Energy, Politecnico di Torino, Turin, Italy

Danilo Palazzo School of Planning, University of Cincinnati, Cincinnati, OH, USA

Philip D. Plowright College of Architecture and Design, Lawrence Technological University, Southfield, MI, USA

Joaquín Sabaté Bel Urbanism and Regional Design, Polytechnic University of Catalonia, Barcelona, Spain

Angioletta Voghera Interuniversity Department of Regional and Urban Studies and Planning, Politecnico di Torino, Turin, Italy

About the Editors

Roberta Ingaramo Architect, Ph.D. in Architecture and Building Design, is assistant professor in architecture and urban design, Department of Architecture and Design (DAD), Politecnico di Torino, and has a master's degree in Conservation of historic towns and buildings, Katholieke Universiteit, Belgium. She is member of the Scientific Committee of S3+Lab Urban Sustainability & Security Laboratory for Social Challenges at Politecnico di Torino and is visiting researcher in 2016 at CMU in Pittsburgh and invited lecturer at Lawrence Tech (2016) and speaker at ETSAM in Madrid (2015). She is responsible for a research fellowship in a European project and tutor in several ASP (Alta Scuola Politecnica, involving the best students of the Politecnico of Torino and Milano) projects and workshops. Her research deals with architectural and urban design, with specific attention to the different scales involved, urban regeneration, and adaptive reuse of sites and disused buildings also through applied research and award-winning international design competitions. Since 2013, she is coordinator of the public works contracts commission of the Council of Architects, Planners, Landscapers and Conservators of Turin Province. She is author of several articles, book chapters, and two monographs: *Dall'immagine del costruito alla costruzione del progetto* (Celid 2009) and, with Angioletta Voghera, *Planning and Architecture. Searching for an Approach* (Alinea International 2012).

Angioletta Voghera Ph.D. in Regional Planning and Local Development is associate professor of urban and regional planning, Interuniversity Department of Regional and Urban Studies and Planning (DIST), Politecnico of Torino. In the Politecnico di Torino, she represents the "Urban and Regional Planning Area" in the Task Force on Technology Transfer; she is member of the Scientific Committee of S3+LAB-Urban Sustainability & Security Laboratory for Social Challenges and of the South China-Torino Collaboration Lab (Politecnico di Torino-South China University of Technology). In 2014, she was visiting researcher in EPFL of Lausanne. She is the coordinator of the National Commission on Environment and Landscape Planning of the National Institute of Urban Planning (INU), www.inu.it (2014). She is member of the editorials boards of the *American Journal of Civil Engineering*, *Town and*

Country Planning, *GeoProgress Journal*, *Sentieri Urbani: Rivista dell'INU Trentino*. Participating in territorial governance and elective organization, since 2013 she is member of the Council of Architects, Planners, Landscapers and Conservators of Turin Province and since 2016 of the Administration Council of Parchi Reali (Piedmont Region). Resulting in over 160 international and national publications, her main research activities are in the field of regional and landscape planning and urban design focusing on analysis, evaluation, planning, and design of sensible environments, such as natural and landscape protected areas and fluvial and rural landscapes. The main publications in these fields are the monographs *Culture Europee di Sostenibilità: Storie e Innovazioni Nella Pianificazione* (Gangemi Editore 2006), *After the European Landscape Convention: Policies, Plans and Evaluation* (Alinea 2011), and with Ingaramo R. *Planning and Architecture: Searching for an Approach* (Alinea International 2012). She is editor of *Progetti per il Paesaggio* (Inu Edizioni 2015) and – with C. Bianchetti, E. Cogato Lanza, A. Enver Kercuku, and A. Sampieri – *Territories in Crisis: Architecture and Urbanism Facing Changes in Europe* (JOVIS Verlag GmbH 2015).

Abbreviations

A	Albedo coefficient
ANP	Analytic network process
ARPA	Regional Agency for the Protection of the Environment of Piedmont Region
ASTER	Advanced Spaceborne Thermal Emission and Reflection Radiometer
BCR	Building coverage ratio
BD	Buildings' density
BKP	Beeldkwaliteitplan
CoE	Council of Europe
CSI	Consortium for Information System of Piedmont Region
DA	District authority
DN	Digital numbers
GI	Green infrastructure
GIS	Geographic information system
HDD	Heating degree days
ISPRA	Istituto Superiore per la protezione e la ricerca ambientale
IUCN	International Union for Conservation of Nature
LCA	Landscape character assessment
LDF	Local development framework
LST	Land and temperature surface
LULC	Land use and type of coverage
MOS	Main street orientation
NDVI	Normalized vegetation index
NEPA	National Environmental Policy Act
P	Solar Exposure factor
PLU	Plan local d'urbanisme
RC	River contract
SCoT	Schéma de Cohérence Territorial
SEA	Strategic environmental assessment
SSP	Special strategic project
SWIR	Shortwave infrared

SWOT	Strengths, weaknesses, opportunities, threats analysis
TOA	Top of atmosphere
U	Urban morphology factor
UHI	Urban heat island
UN	United Nations
UNESCO	United Nations Educational, Scientific and Cultural Organization
UPC	Catalonian Polytechnic University
W	Average buildings’ distance

Chapter 1
From the River to the Project

Roberta Ingaramo and Angioletta Voghera

Keywords Territorial design • Landscape design • Urban and architectural design

The city and territory are the outcome of a political project that develops in space (Geddes 1904, p 113). Every society spreads from structures and infrastructures that are necessary to the life of an urban settlement, conveying shape to space, controlling it and arranging it to become organized.

The role of design in forming cities, territories and landscapes is central and is the result of political, theoretical, methodological and formal intentions which become rooted and transform in time and space "using spatial planning as their main technical tool" (Mazza 2008). This technique articulates the aims of the political project in spatial terms, redesigning the forms of citizenship (Mazza 2011, 2015).

Morphological design interprets and practices this and is essential to the structuring of change, constituting the cognitive, argumentative and conservative framework for the definition of incremental strategies. The results of design are not merely the end, but also the means for carrying out transformations (Palazzo 2008; see also Chap. 3, Ingaramo and Voghera 2012, De Rossi et al.).

The role of the architectural and urban project "as a producer of knowledge", broken down by Paola Viganò (2010, p 11) into three theories open to different developments and articulations, is characterized by conceptual elaborations: abstractions relating to a space and time which make use of multidisciplinary knowledge; descriptions as the recognition of situations, players, places, their relationships, orders and rules; and sequences of actions, decisions dislocated in time or

R. Ingaramo (✉)
Department of Architecture and Design, Politecnico di Torino,
Viale Mattioli 39, 10125 Turin, Italy
e-mail: roberta.ingaramo@polito.it

A. Voghera
Interuniversity Department of Regional and Urban Studies and Planning,
Politecnico di Torino, Viale Mattioli 39, 10125 Turin, Italy
e-mail: angioletta.voghera@polito.it

R. Ingaramo, A. Voghera (eds.), *Topics and Methods for Urban and Landscape Design*, Urban and Landscape Perspectives 19,
DOI 10.1007/978-3-319-51535-9_1

scenarios, which explore relationships between different theories of spatial transformation. The various roles of design as an act of proactive knowledge are based on the idea that it is an open process, in which the various socio-economic and institutional actors and the different positions at stake interact to present new stances, developments and outlooks on the future of space.

Design strongly expresses its epistemological value, which sees practices not as applications but as tools for reflection and research (Schön 1983), taking on considerable social value, particularly because it becomes an opportunity for comparison, assessment and interpretation of people's hopes (European Landscape Convention 2000).

This is the research approach taken by this book, which analyses the role of design in people's living space, starting with methodological experiments (see Part I "Within the Design Experimentation" and Part III "Debate") and design practices (see Part II "Case Studies") which assign it the important job of building up interpretations of the territory and future visions of the landscape on different scales, investing the city, architecture, nature and agriculture, in their reciprocal relationships.

Interpretation and assessment are nodal aspects of the gradual construction of design as a territorial innovation process, defined in conjunction with institutional and social stakeholders (see Chap. 5, Brunetta).

In fact, the interpretative and assessment frameworks supply indications for a transcalar design which seeks coherence between large scale and small scale (Berger 2008, Waldheim 2006), binding policies, landscape projects and plans at urban, local and spot level by effectiveness (see Chaps. 2, Voghera and 3, Ingaramo); these designs (integrated, strategic, etc.) are capable of a transcalar overview, paying attention to the environmental and socio-economic dimensions of landscaping processes, which are always necessary when planning landscapes, and a prerequisite for launching synergic top-down and bottom-up actions, as required by the ELC (Voghera 2015).

1.1 River Contracts: Suggestions for Territorial and Landscape Design

The research starts with the river area, a place where the effects of climate change have a stronger influence and where design has to generate resilience (Sendzimir et al. 2007, Cosens and Williams 2012); besides, there are territories where – thanks partly to the river contract method – design experiments open up the way to processes of knowledge, reclamation and governance, with the help of social stakeholders (McCluney et al. 2014).

Rivers act as corridors contributing to the health of urban ecosystems that can be described as a connection of the patches of the "landscape mosaic" (Forman 1995, p 209) in a dynamic equilibrium of hydrologic flows, animal and natural dynamics and human activities (e.g. the interaction among human and natural activities for urban planning in Chap. 11, Mutani and Fiermonte), each one influencing the others.

With this in mind, it is strategic to develop a new perspective for action along the rivers based on inclusiveness and multilevel governance, in order to implement the link between spatial policy and urban and landscape design. Brunetta (Chap. 5) shows the innovative vision in the Italian context that goes beyond the traditional sectorial approach, aimed at promoting the coordinated, synergic action of different policies regarding water, protection of the land, biodiversity and landscape. The new method is based on the experimentation underway along the River Po, which configures an inseparable system of relationships between actions, objects and stakeholders linked to the territory and landscape. The policies promoted by the Po River Basin Authority encompass not only sectoral planning actions but also uses and activities of the territory, involving territorial institutions, park authorities, other sector representatives and, in particular, the local populations, who become actors and must be informed and involved in the discussion, debating and sharing of a complex urban project, the success of which is closely tied to joint participation in decisions and their application.

According to this approach of organizing and enhancing territories and landscapes at basin scale, river contracts – as experienced in Italy, France and Belgium – may be useful tools to restore both ecosystems and the sense of community at local and regional level; value enhancement processes affect very different territories, from the urban environment to the countryside and natural context, promoting the evolution of socioecological systems and their landscapes on a regional scale (Chap. 2, Voghera).

With this in mind, the contribution by Berruti and Moccia (Chap. 6) and Ingaramo and Voghera (Chap. 4) focuses on river contract experimentations as water management tools that require an appropriate multidisciplinary knowledge and streamlined planning process to be constructed and launched. The river contract is an appropriate method, useful in the development of a cognitive framework for the arrangement of action plans and the involvement of all the relevant stakeholders. The chapter describes simulations of different stages of river contract processes carried out during urban planning courses in the Urban, Landscape and Environmental Planning Program, at Federico II University of Naples. This didactical experimentation starts from the knowledge-based activities and defines some strategic actions that are a common framework of river landscape transformation (see Chap. 2, Voghera; Chap. 3, Ingaramo; Chap. 5, Brunetta; Chap. 8, Sabaté Bel; Chap. 7, Mellano; Chap. 10, De Rossi et al.) in order to:

- Recover river landscapes (re-naturalization, restoration, reconversion, reuse).
- Develop "dynamic landscapes" as a specific way to enforce the "relationship between resilience and resistance and pose the question of how to redesign the borders between spaces able to adapt to the presence of water and spaces destined to remain unchanged in the course of time" (such as in agricultural areas affected by strong hydraulic risk, the controlled flooding green areas, the redesign and re-naturalization of river banks together with a storm water management approach).

- Enforce riverfront fruition dealing with the natural functioning of river corridors and their capacity for activating liveable landscapes.

Starting with river contracts, as described in Voghera (Chap. 2), territorial and landscape design now has new meanings and roles, referring to systemic relationships in the wide areas involved, the society and actors responsible for transforming and managing the landscape and the different resources (natural, landscape and socio-economic) included in the development scenario. It is a project that links visions, expressing utopias and effective actions (Magnaghi 2005); this project tries to achieve a balance among ecology, culture and aesthetics, but it also incorporates strategic and operational tools with reference to the broad debate on strategic planning, which assigns design different forms and roles for the transformation of the territory in relation to urban and regional planning.

Within the planning framework, this conception of design plays a key role in the choices of action spaces and in the definition of action times, suggesting hypothetical transformations in a general durable and locally shared framework; focused on the action in a specific area, the traditional conception of the development plan (master plan in the recent common use) is assigned to smaller or local designs. In the discussion on strategic planning, this is a very general and wide-scale vision, implemented through operational projects; the strategy indicates constraints and limits, giving the systemic vision a strategic value, as an overall frame of each transformation.

Indeed, the project plays a programmatic role in the identification of a system of values for the future of a territory, indicating a coherent framework for areas of action, referring to different action modalities and, sometimes, indicating priorities and responsibilities for implementation, as in the Stura and Sangone River Contract Master Plans and related specific projects (see Chap. 3, Ingaramo and Voghera).

Discussing some examples, like the river contracts in some European countries, and case studies (Part II), territorial and landscape design becomes a tool for promoting cooperative and competitive collaboration among territories, different tools and diverse social actors according to a "place-based" development approach (Barca 2011). It is a project with a transcalar and multidisciplinary and sustainable method, evaluated through an extensive negotiation process for sharing decisions, which sometimes includes participation (at least stakeholder perspectives in Chap. 5, Brunetta).

With this in mind, the design experience of the Llobregat River Heritage Axis Master Plan (VV. AA 2001, MIT, UPC, Chap. 8, Sabaté Bel) provides, thanks to official recognition by UNESCO, a path for future growth of the colonial river landscape that is slowly attracting private economic expansion. On a territorial scale, it promotes links between colonies, doors and also museum and commercial nodes of the proposed heritage park. On an urban scale, colonies constitute an active and complex heritage in the relationship with buildings, open spaces and the river landscape. The design of this cultural system needs cooperation to ensure the success of a heritage park, starting from recognizing the role of bottom-up contribution and the value of the local community and residents in its promotion and design. Integrating different scales and actors, cultural landscapes are "capable of bringing together vast regions and scales of activity, transcending the limits of a specific level of government".

So design (as indicated in Chap. 2, Voghera and Chap. 4, Ingaramo and Voghera) is not only a technical and/or institutional action but a series of specific actions, strategies, guidelines and participatory events connected to each other and placed in a dense field of social and institutional policies and practices. Design is an open and multidimensional practice, characterized by interacting technical and political outcomes (not necessarily cooperative and in many cases conflicting).

So the design activity takes on the form of a system of actions, practices and interventions, but also of not necessarily intentional outcomes. In this activity all the actors involved play a central role in the process through which aspirations and interests of local communities interact, involving local people, institutions and public and private actors (ELC 2000).

The interpretive discussion of different cases in the book helps to indicate ways of integrating strategies and operational actions for the development of the territorial values, enhancing the available assets and promoting sustainable development consistent with local resources, with identity of the place and with local expectations.

In this way, landscape design mobilizes a multiplicity of resources and actors around a "common vision". The project aims, therefore, address the individual actions in an organic system, identifying possible forms of coordination, management and financing tools for implementation.

The examples indicate the consolidation of a development planning method based on a variety of independent projects of small scale (local projects), developed as part of a larger project (project framework) (Mazza 1999 and 2004, Campos Venuti 1994, Faludi and van der Valk 1994, Gregotti 2000), where the landscape is central to promote quality of life; this approach requires a large-scale project which consists of a long-term, flexible and adaptable multitude of small-scale/local designs aimed at capitalizing, exploiting and managing resources, with medium- or long-term effects (see Chap. 2, Voghera).

The impact of design depends on the capacity of the long-term or occasional policies to support territorial governance and management, defining innovative tools for implementation (i.e. Action Contract in Ingaramo and Voghera 2012, Chap. 4, Ingaramo and Voghera) starting from the transcalar role of architecture (Chap. 3, Ingaramo).

1.2 Transcalar Architecture

The river, a linear morphological object (Chap. 10, De Rossi et al.), can be paradoxically interpreted as a horizontal "zero cubature" object which, according to Aymonino, is "actually a heterogeneous subject which brings together different disciplines and anthropic conditions, with intermediate (or other) conceptual scales between the Plan and architectural object, mediation for the rediscovery of the ethic nature of architecture", which can provide a social service for the design of territories and landscapes (Aymonino 2012).

Too often, transcalarity and interdisciplinarity are rhetorical references to define design strategies from landscape scale to urban and architectural scale (Chap. 3, Ingaramo). In most cases, the definition of design strategies on a large scale does not implement into urban planning and design on the local scale of city and architecture (see Marseille EuroMéditerranée). A charming master plan or a fascinating strategic scenario gives the illusion of the integration of a system, which will probably be developed in its individual parts through an allocation of building lots to investors, who will commission different architects who will then propose their own personal vision of architecture, forgetting the common strategy (Chap. 3, Ingaramo).

In the current economic condition, which sees a continuing lack of public resources, the wide-ranging strategies for landscape and urban design, developed in periods of strong growth, should be reviewed, looking at current strategies such as urban regeneration and adaptive reuse of industrial complexes with pollution, also at landscape level. These approaches require more minute and sensitive attention to individual fragments of urbanity, where even small and subsidiary interventions capable of triggering processes of redefinition of the urban tissue are identified (Chaps. 7, Mellano and 9, Manzo).

If it is true, as Bauman (Bauman and Bordoni 2015) argues, that the stagnation of the economic crisis we have been experiencing for several years is the result of a divorce between power and politics, it is clear that the state's inability to manage the change, in Italy more than anywhere else, has resulted in a stabilization of the crisis that some scientists, like Bordoni, believe can continue forever.

Architectural and urban design, renewed in their strategies, may still be capable of inducing some form of change, capable of triggering regeneration processes with long-term benefits. Looking to architectural and urban design approaches, intermediate places, space in-between by Aldo van Eyck (Van Eyck 1963; Strauven and Van Eyck 1997) and threshold by the Smithson (1968), reviewed in present times, can be the sharing place where transcalar design scenarios can be implemented at landscape and local scale. Traditional zones (Moyo in Ministry of Housing Spain (2010)) are spaces of mediation between built and unbuilt, where we can experiment with transformation projects and processes. Time is the main factor in processes of transformation, but in the age of Twitter, there are possible strategies of intervention in the city that need long implementation times and have to be built in unconventional ways, with low-cost investment, even designed for temporary uses. Participated design processes, although limited to some of the users or stakeholders involved, take on a role of political legitimacy of public decisions when administrations and governments have difficulties in defending them, especially if the choices made are imposed from above.

Our experience in recent years in voluntary involvement in the processes of governance of river contracts in Italy has shown that the identification of expected landscape and local projects, proposed with alternative scenarios, has enabled the definition of shared strategies at the different levels involved (Chap. 4, Ingaramo and Voghera).

Design builds up an incremental transformation scenario which is born locally (as in the case of the Sangone river park and Gesso and Stura riverside park eco-network strategic project; Chap. 10, De Rossi et al.) and acquired "narrative" potential, a recognitive and reconstructive opportunity for the landscape. It defines a landscape vision known as (g)rand(a)stad (De Rossi 2009), a political image of large-scale architecture or, better, of the role that architecture can play in territorial and landscape design: building the development of the city and the territory strongly characterized by dispersed and polycentric elements, starting with local landscape and cultural identity.

As Rossi wrote in the 1960s, architecture represents an aspect of a more complex reality, of a particular structure, but at the same time, being the last ascertainable element, it constitutes the more tangible point of view with which to address the problem of the city. If this is still true, and if Italian and European cities are still a model of reference for urban design, tools and methods of urban design should be reviewed with attention to the historicizing of our urban tissues, focusing attention on this *space in-between*, public, semipublic and private, where changes can be implemented for reconstruction or new foundation of cities.

1.3 Utopias and Methods

Francoise Choay (2005) remembers the considerable influence of utopian visions for architectural and urban design in *L'utopie et le statut anthropologique de l'espace édifié*. Utopia and visionary thinking in planning and design have historically been strong avenues to confront the political, social and environmental adversities of society. As we are reminded by Palazzo and Diko (Chap. 12), the challenges of urban pollution, natural resource depletion, water and air quality and the abominable living conditions of the urban population after the advent and the consolidation of the Industrial Age all inspired new proposals of urban visioning and planning. Ecological planning, particularly in the USA, took a utopian vision towards solving problems or adapting solutions to balance social, economic, political, spatial, environmental, institutional and cultural units. Accordingly, utopia also promoted social transformation by ensuring equitable access to economic and productive resources, creating harmony between town and country. This vision defined holistic approaches that are now applied in ecological planning, and it is the common approach taken to design along the river. For the future of cities and settlements, utopian and visionary thinking-based planning and design continue to be an issue, and it is important to manage existing challenges by implementing strategic adaptation actions.

In "good" planning and design, design methods are not addressed with the same clarity and the same cognitive processes found in other disciplines (Chap. 13, Plowright). In fact, in design, we apply what is called a philosophical position, starting bias or a belief structure. Besides, we choose the goals or objectives that too are part of design method identification, and they span disciplinary boundaries. Both mission statement and goal identification occur before the design method, as part of the setting up of a starting state. Plowright (2014) says that the values from these design positions are needed to frame the testing structure that is used to identify which decisions should be made and why. Moving into a design method, the strategy is the general plan of action, the adaptation of a framework to an explicit design method. Once a strategy (method approach) has been determined, then tactics are developed with cognitive processes based on convergent-divergent thinking in interdisciplinary discussion. Scaling methods are about transferring values between missions and goals to strategies and tactics that occur within a design method. Disciplinary bias is then aligned with overall values to ensure a larger degree of coherence between expected outcomes and disciplinary delivery.

According to Plowright's theory, frameworks and methods can be thought "as a sequence of containers for applied action tools". So, based on a design framework linking social and ecological values, methods immediately narrow options and starting position, creating a series of limits rather than being open to any influence. This is important in a complex system to understand approaching the design process with a design framework, a set of biases and a value system. Besides, without a guiding agreement that spans disciplinary values, design choices will gravitate towards those more naturally found in the historic values of the operating discipline, be it architecture, urban and regional planning, biology, agriculture or politics.

References

Aymonino A (2012) Più spazio meno volume: un racconto in movimento. In: Ricci M (ed) Nuovi Paradigmi. LISt Lab, Trento

Barca F (2011) Alternative approaches to development policy: intersections and divergences. OECD Regional Outlook 2011:215–225

Bauman Z, Bordoni C (2015) Stato di Crisi. Einaudi, Turin

Berger A (2008) Conversazione sul landscape urbanism. In: Viganò P (ed) Landscape of urbanism. Officina Edizioni, Rome

Campos Venuti G (1994) La Terza Generazione dell'urbanistica. Franco Angeli, Milan

Choay F (2005) L'utopie et le statut anthropologique de l'espace édifié, in Esprit, n. 318

CoE (2000) European landscape convention. Council of Europe, Florence

Cosens BA, Williams MK (2012) Resilience and water governance: adaptive governance in the Columbia River basin. Ecol Soc 17(4):3

De Rossi A (ed) (2009) GrandeScala. Architettura, Politica, Forma. List, Barcelona

Faludi A, van der Valk A (1994) Rule and order: Dutch planning doctrine in the twentieth century. Kluwer Academic Publishers, Dordrecht

Forman RTT (1995) Land Mosaics the ecology of landscapes and regions. Cambridge University Press, Cambridge

Geddes P (1904) Civics: as applied sociology. Dodo Press

Gregotti V (2000) Progetto urbano e pianificazione: un dialogo necessario. Urbanistica 104:126–127

Ingaramo R, Voghera A (2012) Planning and architecture. Searching for an approach. Alinea International, Florence

Magnaghi A (2005) The urban village: a charter for democracy and local self-sustainable development. Zed Books, London
Mazza L (1999) Trasformazioni del piano. Franco Angeli, Milan
Mazza L (2004) Piano, progetti, strategie. Franco Angeli, Milan
Mazza L (2008) Geddes 'politico': vision, survey, citizenship. Territorio 45:91–98
Mazza L (2011) Dimensione urbana e strumenti di governo del territorio, 23rd October 2011. Available at: http://www.empirismoeretico.it/luigi-mazza-dimensione-urbana-e-strumenti-di-governo-del-territorio. Accessed 31 Jul 2016
Mazza L (2015) Spazio e cittadinanza. Politica e governo del territorio. Donzelli Editore, Rome
McCluney KE et al (2014) Riverine macrosystems ecology: sensitivity, resistance, and resilience of whole river basins with human alterations. Front Ecol Environ 12(1):48–58
Ministry of Housing Spain (2010) Social Housing & City
Palazzo D (2008) Urban design. Mondadori Università, Milan
Plowright P (2014) Revealing architectural design: methods, frameworks & tools. Routledge, London
Schön D (1983) The reflective practitioner. Basic Book, New York
Sendzimir J, Magnuszewski P, Flachner Z, Balogh P, Molnar G, Sarvari A, Nagy Z (2007) Assessing the resilience of a river management regime: informal learning in a shadow network in the Tisza River Basin. Ecol Soc 13(1):11
Smithson A (ed) (1968) Team 10 primer. New York, Studio Vista & Reinhold
Strauven F, van Eyck A (1997) The shape of relativity. Architecture & Natura, Amsterdam
Van Eyck A (1963) Beyond visibility, about place and occasion, the in between realm, labyrinthian clarity. Situat Times 4:79–85
Viganò P (2010) I territori dell'urbanistica. Il Progetto come produttore di conoscenza. Officina Edizioni, Rome
Voghera A (ed) (2015) Progetti per il paesaggio. Libro in memoria di Attilia Peano. INU Edizioni, Rome
VV.AA (2001) Designing the Llobregat Corridor: cultural landscape and regional development. Catalonian Polytechnic University/Massachusetts Institute of Technology, Barcelona
Waldheim C (2006) The landscape urbanism reader. Princeton Architectural Press, New York

Part I
Within Design Experimentation

Starting from our research and design experiences, the first part of this book reflects on the various roles of design as proactive knowledge, an open decision-making process where situations and places, social and economic actors, strategies and rules interact in spatial transformation (Ingaramo and Voghera). The research analyses the role of design in people's living places (dealing with the European Landscape Convention 2000) constructing incremental and relational design visions to integrate nature, agriculture, city and architecture for a specific river environment.

Territorial and landscape design is the tool for promoting cooperative and competitive collaboration among territories integrating strategic and local policies, different social actors and visions for a location-based scenario.

Chapter 2
Approaches, Tools, Methods and Experiences for Territorial and Landscape Design

Angioletta Voghera

Abstract Territorial and landscape design now has new meanings and roles, referring to systemic relationships in wide areas, the society and actors responsible for landscape transformation and management and the different resources included in the development scenario. It links visions, expressing utopias and effective actions, trying to achieve a balance among ecology, culture and aesthetics and settlement assets, but it also integrates strategic and operational tools with reference to the wide debate on strategic planning. Within this framework, this chapter discusses several international and national experiences – such as the River Contracts or Corona Verde in Piedmont – that use different tools based on trans-scalar, sustainable and sometimes participative methods for promoting cooperative and competitive collaboration between territories, according to a place-based approach.

Keywords River contract • Territorial design • Landscape design • Project governance • Design process • Participation • Environmental compensation plan

2.1 River Contract and Environmental Compensation Plan: Tension Between Utopian and Operational Dimension for Territorial and Landscape Design

A vast debate is in progress on the current pertinence and utopia of the contemporary design of territorial transformation scenarios, involving Francoise Choay in *L'utopie et le statut anthropologique de l'espace édifié* (2005) and Didier Minot in *Le Projet de territoire* (2001), with reference to the important role of "agreements" for the enhancement of the landscape and local development, like the River Contracts in Italy, Belgium and France or the Landscape Contracts, known as *contract de pays* in France and *Contratti di paesaggio* in Italy. To varying extents, in the definition of

---

A. Voghera (✉)
Interuniversity Department of Regional and Urban Studies and Planning, Politecnico di Torino, Viale Mattioli 39, 10125 Turin, Italy
e-mail: angioletta.voghera@polito.it

R. Ingaramo, A. Voghera (eds.), *Topics and Methods for Urban and Landscape Design*, Urban and Landscape Perspectives 19,
DOI 10.1007/978-3-319-51535-9_2

actions to transform the territory, these tools involve the relationship between resources and their use, as well as the role of local communities and landscape for territorial planning and design.

And it is also in relation to the strategic scenarios of ecological and landscape transformation of the riverside in the River Contracts (RC) shared by the communities in relation to a river basin that the reflections of this book are prompted. The RC is a method rather than a territorial governance tool, which helps identify shared strategies, actions and rules for the environmental, landscape and socio-economic requalification of a hydrographic basin. The RC is an experience which has been conceived and developed over the last decade and is being gradually consolidated in methodological and operational terms, thanks to an increasingly widespread recognition of its value in the management of water and the more general planning of the riversides[1] and recent recognition by several regional laws (Piedmont, Lazio, Lombardy and Marche), and by law 221/2015, among the actions to sustain the green economy and guarantee sustainable use of natural resources.[2] Starting with the aim of contributing to the improvement of water quality, implementing directive 60/2000/EC, it is a method that is being reinforced in the construction of a project capable of enhancing the territory and landscape from a strategic, multidisciplinary and trans-scalar viewpoint, defining systems of action at basin level and "detailed" projects that can help implement or launch innovations in different territorial plans, starting with the basin.

The RC promotes vertical or horizontal subsidiarity; interactive local development; the safety of the riverside and sustainability (National Charter of River Contracts, V National Table on 21 October 2010 in Milan), via a decision-making process that aims to include those involved and integrate the themes undertaken (Carter 2007); unhinging the traditional forms of water governance based on top-down hierarchies and overcoming the strictly technical and sectoral nature (Eckerberg and Joas 2004).

The RC, created on the basis of strategic and negotiated planning, starts with a voluntary agreement and mobilises participation by all the stakeholders along the riverside for the concerting of conflicts and the definition and implementation of a general and shared riverside development framework (Affeltranger and Lasserre 2003) (Fig. 2.1).

---

[1] The River Contract is now increasingly recognised in the form of different tools: basin plan or hydrographic district, water protection plan, landscape plan, regional and provincial territorial plan and rural development plan (Bastiani 2012).

[2] Law no. 221, dated 28 December 2015, contains measures for the protection of nature and sustainable development, environmental assessments, energy, green purchases, waste management and reclamations, soil protection and water resources. The provisions to guarantee universal access to water include the river agreement, which is recognised as a voluntary tool of strategic and negotiated planning for protection, the correct management of water resources, the safety, protection, management and enhancement of riversides and local development, contributing to the definition and implementation of tools for district planning at hydrographic basin and subbasin level (art. 59).

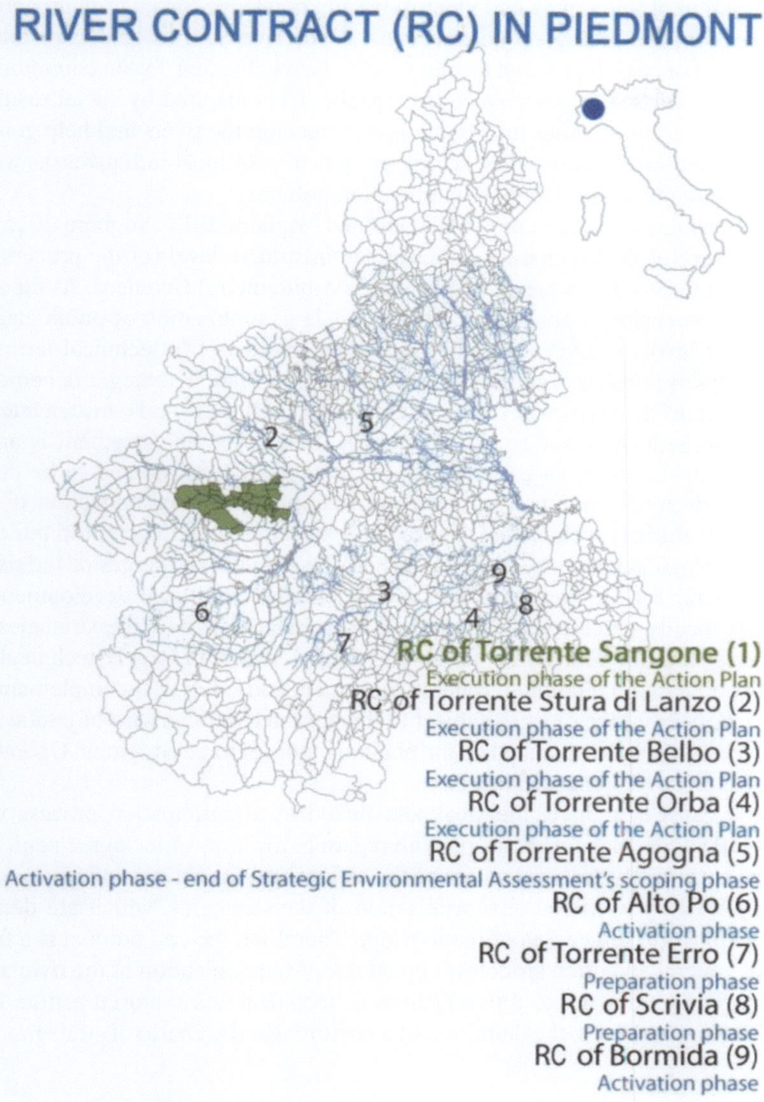

Fig. 2.1 River Contracts in the Piedmont Region (Source: Author's elaboration)

Consequently, the decision-making process should involve as many different stakeholders as possible, both in terms of socio-economic nature and of importance within the decisional arena. The aim is to create a territorial design with integrated broad contents (soil and water protection, environmental improvement, landscape enhancement and territorial development) and integrated forms of funding, to orient planning, programmes and policies. With this in mind, it can also contribute to the reconstruction of know-how and identity-based knowledge linked to hydrogeological protection, the ecological enhancement of the river and its landscape and the development of multifunctional farming methods, reactivating "basin communities" in forms of inter-scalar governance (Magnaghi 2011) inspired by social resilience (Pearce 2003). This enables the populations to reclaim the rivers and help generate new urban and rural territoriality, setting up networks of local initiatives, as well as generating integrated territorial requalification policies.

In analysing the Italian cases (Avidano and Voghera 2012; Voghera 2015), we can trace several shared characteristics at administrative level (of the process) and with regard to project and technical content, despite their differences. As far as the concerted dimension is concerned, the process is a combination of public and private, and it involves a very varied range of stakeholders.[3] In technical terms, the cases analysed reveal a predominance in the Action Plan, a strategic document of the Contract, of the themes linked to water and soil pollution, and constant attention to the hydrogeological safety of the territory. Landscape enhancement is an aim shared by all the different experiences, but is generally considered to be closely linked to fruition, deriving from the construction or the interconnection of slow mobility itineraries (cycle paths, panoramic roads), and to the attempt to build portions of ecological network on a local scale. While a vision emerges of the river as a nodal factor for the quality of life and flywheel of territorial development, but often not locally perceived or considered as a threat to safety, the strategies proposed refer too often to ordinary planning actions and with largely technical contents (particularly in relation to water management and safety). The implementation of the actions envisaged is also subject to the presence or otherwise of public funds that have already been allocated by the planning tools in force (Regional Operational Plans, Rural Development Plans, etc.).

In the cases examined, the weakness linked to a participation process which makes little allowance for inclusion with regard to the aims of the agreement, along with the excessively sectorial nature of the actions and the absence of specific funding, is joined by a limited territorialisation of the strategies, which are detached from the physical dimension of the territory. Therefore, the end product is a framework of general and often spaceless objectives, of requalification of the river and its territory, broken down into different lines of technical and sectorial action. In the experiences considered, the definition of a territorialised scenario of strategies is not

[3] While the organisational structure is very different from case to case, the presence of a small decision-making body (Direction) made up of the most important subjects emerges. This body coordinates the activities of the agreement and outlines the strategic orientations. A second body (Basin Meeting, Agreement Forum) tends to be made up of all the subjects that, in different ways, become part of the process. It is usually defined as the extended participation body, although the main tendency is to include the stakeholders with more weight – from the big economic players to the institutions and representative associations according to different models of participation (workshops, focus groups, meetings, etc.). In terms of participation, certain experiences have ignored every form of consultation of the public; others have involved only certain age groups (mainly through experiences with schools), while others again have considered only specific groups of subjects (stakeholders).

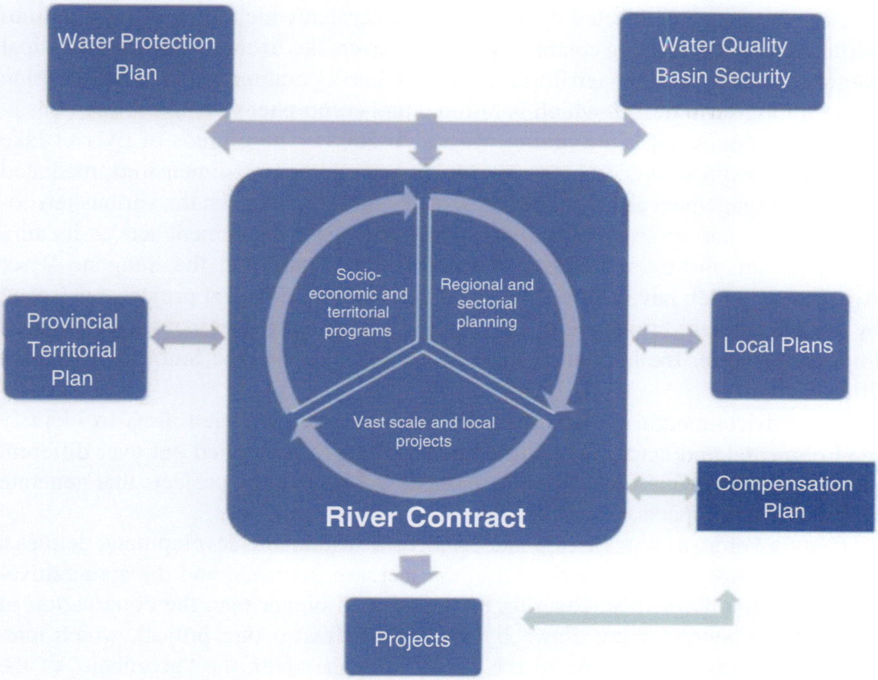

Fig. 2.2 The River Agreement (Source: Author's elaboration)

usually reached in the form of a "territorial project", which could allow vision of the "spatial" outcomes, guiding the process of revision and implementation of the agreement (Fig. 2.2) and also revealing the physical and functional interactions between the various interventions envisaged.

There are a few exceptions among the Italian RC experiences, like those experimented in Piedmont (Fig. 2.1), which are implemented by combining the strategic apparatus of the Agreement – the Action Plan – with a Masterplan which orients the vast area territorial action and with the preparation of an action agreement, which guides local projects, creating an agreement for cooperation in terms of methods, criteria and financial resources between public entities and private players involved in the project, with the aim of guiding the single transformations (Ingaramo and Voghera 2012a). This is definitely the case of the first River Contract in Piedmont (Contratto di fiume del Torrente Sangone, signed in 2008) – that of the Torrente Sangone, which opened up the way to the experimentation of a method, partly repeated for the Stura di Lanzo and for the lakes of Avigliana and Viverone – in which certain operations envisaged by the Masterplan (Ingaramo and Voghera 2012b, promoted scenarios for the future of the territory, suspended between utopian visions and operational actions: limitation of soil consumption, redefinition of the limits between town and country, enhancement of the networks and landscape systems with the recovery and integration of the main historical-cultural and symbolic focuses.

The visions are connected by a general Masterplan which transforms them into territorial projects, creating conditions along the riversides to coordinate the municipal authorities and put other territorial strategies into operation too. The Masterplan offers a long-term design which is rational and comprehensive, to convey consistency and communicate the future (Magnaghi 2005). These kinds of river or lake agreements express tension between utopian and operational dimensions, mediated by social engagement and the construction of synergies between the various territorial planning and programming tools, which the River Agreement sets as its aim. Examples are the interrelations between Corona Verde and the Sangone River Agreement, which have allowed the accomplishment of several projects envisaged by the Masterplan; besides, the Stura river Environmental Compensation Plan has important role for the implementation of actions related to the Stura RC (Voghera 2014–2016).[4]

The Environmental Compensation Plan is a method which aims to identify environmental and landscape enhancement actions to be carried out over different periods of time using funds from the compensations paid by projects that generate environmental impacts.

Corona Verde, a strategic regional project for sustainable development, defines a strategy to improve the quality of life, the increase in appeal and the competitiveness of 93 municipalities, covering an area that is bigger than the conurbation of Turin. The boundaries are drawn by the green infrastructure project, which integrates the *Corona di Delizie* of the royal residences with the "greenbelt" of the metropolitan parks, rivers and rural areas. It promotes integrated ecological and landscape reorganisation and requalification, implementing a long process of cooperation. The projects become part of a system, especially in territories where the municipalities work together on the RC (Stura and Sangone); in general, the Masterplans of the other four territories (out of a total six territorial areas)[5] into which the Corona Verde area has been broken down (Fig. 2.3) were based on the need to present a complete Masterplan within the short timescale defined by Regione Piemonte, which coordinates the project and the allocation of funds, linking different local projects which had been put on the backburner by the local administrations for accomplishment as and when resources became available.

Corona Verde has definitely given the projects envisaged by the RC Action Plan financial independence. Originally it was intended to promote the ecologic functionality of the territory for the active conservation of natural spaces and networks, the defence of the hydrographic grid, the defence of rural space in marginal and border areas as well as the enhancement of historical-cultural heritage and the

[4] Research in progress with the Metropolitan City of Turin in relation to the definition of a method to guide environmental compensations and build the ecological network along the Stura river and in close connection with the River Agreement.

[5] Northwest area Masterplan – Lead Municipality Venaria Reale; North area Masterplan – Lead Municipality Settimo Torinese; East area Masterplan – Lead Municipality Chieri; South area Masterplan – Lead Municipality Nichelino; West area Masterplan – Lead Municipality Rivoli; Torino Masterplan.

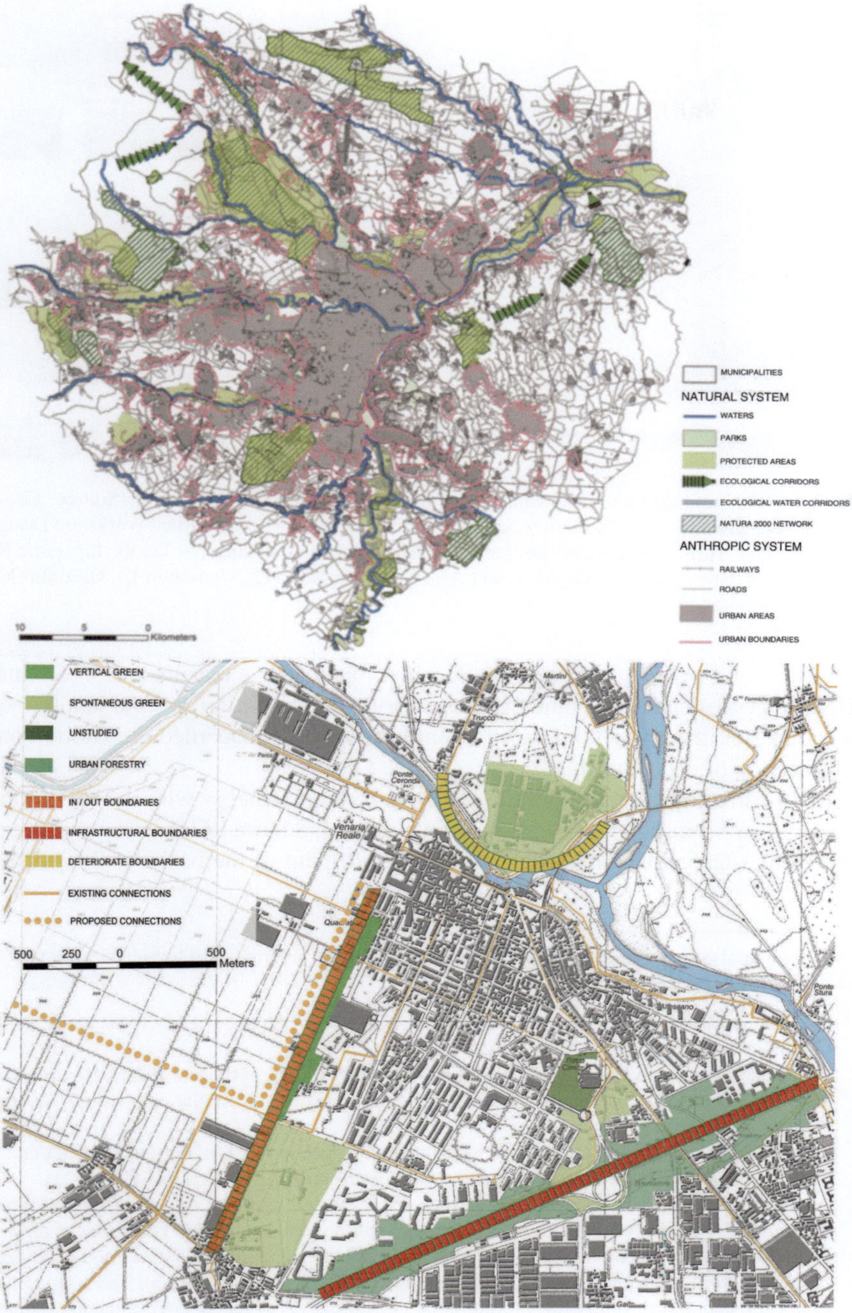

Fig. 2.3 The Masterplan of the urban boundaries and project strategies in Venaria (Source: Basile F., Dissertation on Territorial, landscape and environmental planning, DIST, Politecnico di Torino, Strategy of the green boundaries of Corona Verde. A project for the enhancement of the marginal landscape between town and country, between river and natural, rural areas. Supervisor Voghera A., co-supervisors: Chiantore D. (Regione Piemonte) and Devecchi M.)

Fig. 2.4 Re-naturalisation and reclamation of the quarry along the Stura river (Source: *Green Line* – river, networks, park/*Green line*, una rete tra fiume e parco – International Workshop Design along the river, Turin – September–December 2014 – Politecnico di Torino – Tutors: Ingaramo R., Voghera A., Claps P., Mutani G., Mazzotta A. – Students: Bello C., Comoglio E., Destudio M., Nuzzo F., Sandrone I.)

environment, weaving strong relationships between elements of excellence (UNESCO sites and assets) and the widespread system of rural cultural landscapes. In actual fact, the projects financed so far have mainly concerned the fruition and construction of cycle paths.

In this direction, along the Stura di Lanzo river, in synergy with the RC, I have proposed an environmental compensation plan[6] of the Stura RC territory capable of promoting the ecological and landscape enhancement of the basin, with systemic

---

6 Environmental compensation is a theme discussed widely by international literature, and some of the most consolidated stances are listed here:

- "the provision of positive environmental measures to correct, balance or otherwise atone for the loss of environmental re source" (Cowell 2003).
- "the creation of new values, which are equal to the lost values. If the lost values are irreplaceable, compensation concerns the creation of values which are as similar as possible" (Kuiper 1997)
- "equalising the loss or increasing the environmental values in the proximity of an area that has experienced loss of environmental capital due to development" (Pettersson 2004).
- "measures taken to make up for the loss of, or permanent damage to, biological resources through the provision of replacement areas. Any replacement area should be similar to or, with appropriate management, have the ability to reproduce the ecological functions and conditions of those biological resources that have been lost or damaged" (CIEEM 2016). These guidelines reveal the ecological dimension of combined compensation to repair environmental damage (primarily soil consumption), the operations envisaged by environmental compensation cannot be monetized, and the site where the compensation is carried out is alternative to the project site.

and detailed actions for the re-naturalisation of rundown areas which could become priority nodes for the implementation of the ecological network.

The plan (Fig. 2.4) will be ratified by the municipalities within the Stura river basin and shared during the participated project for the construction of the RC Action Plan. Only via the construction of a compensation plan can we avoid "spot" compensation operations, separate from a strategy of protection, enhancement and restoration of the environment (Cowell 1996) and, therefore, incapable of producing real benefits for the environment and the landscape.

The method on which the plan is based is inspired by techniques to assess compensation operations, such as "No net loss of ecological values" from the Netherlands Rural Development Plan (2004), the Bavarian Ökokonto or the US eco-credits of the mitigation bank. The method develops with the recognition, assessment and classification of the ecological importance of the areas involved in compensation; the definition of the possible compensation measures to improve the ecological and landscape value of the areas of intervention, the definition of intervention priorities and the monitoring and update of the naturalistic value of the areas in which compensation occurs.

A system of areas to be transformed with environmental and landscape interventions is defined, acquiring economic resources from compensations, involving the municipalities engaged in the River Agreement: one of the priority areas is the abandoned quarry at the confluence between the Torrente Ceronda and the Stura river in Venaria Reale, for which reclamation, re-naturalisation and the use as a natural swimming pool, with water purified using plant-based techniques, have been proposed.

2.2 Roles of Territorial and Landscape Design

Thanks to the agreement tools, the territorial and landscape project becomes an instrument for promoting cooperative and competitive collaboration between territories, integrating different tools, according to a place-based development approach (Barca 2011). It is a project with a trans-scalar and multidisciplinary and sustainable method, evaluated through an extensive negotiation process for solving conflicts and sharing decisions based on public participation.

The territory seen in this perspective is the space of action of the urban and landscape design which works to solve problems and create opportunities, governing and coordinating transformative actions (Kunzmann 2007) on different scales in "terms of capacity" of relations in time and in space (Albrechts 2003; Healey 2003). It is not a neutral space where independent events and actions take place, but a space for interaction between complex dynamics which involve the social, economic,

Consequently, compensation projects are an additional resource to qualify the environment, not the result of a nature project made up of isolated pieces, but an integral part of a strategy underlying a precise idea of territory, environment and landscape.

environmental and landscape system. On one hand, it is the space for expressing and comparing political-administrative skills, which interpret the coagulation of values and resources, of uses and representations. On the other, it is the space used, inhabited, known and cared for by those who live in it (Magnaghi 2005), who aspire to the safety, quality and wellbeing of the territory. It is a stratified, "sedimented" space, rich in signs and durable values, the outcome of successive cycles of civilisation linked together in a dynamic process of selective accumulation (Magnaghi 2010), but also of emerging organisational principles, evolutionary tendencies, project problems and possibilities (Lanzani 1991, 2003), as well as practices in terms of the use and mobility that the project indicates for the future (Crosta 2010), combining local community hopes and policies (Bohl and Plater-Zyberk 2006).

Inevitably polysemous and ambiguous, the "territorial" project (Russo 2014), which has been covered widely in urbanistic studies and also in the debate on strategic planning, is gaining new meanings with reference to the crisis. This crisis imposes the need to rethink the consolidated ways of acting, which are often unable to create long-term development (Voghera and Regis 2016). It is, above all, the progressive reduction in public spending that, feeding uncertainty on the orientation of policies, focuses on the endogenous and local capacity to contribute to the processes of change (Giaccaria et al. 2013).

In this direction, RCs are a real opportunity for building the territorial project with the contribution of the communities, to launch that environmental and landscape regeneration, with the transformation of systems of actions, some of which small and molecular, sustained by local actors. This project requires interconnection in an overall strategy, capable of building up through environmental reclamation and regeneration processes and actions which might even be minimal; it also requires considerable effort to create the economic conditions for accomplishment (Lanzani 2015).

Despite the necessary transformations linked to the current crisis, the role of the territorial project in the governance of spatial transformations assigned in the negotiations on strategic planning (Davoudi and Strange 2008) continues to be worth something, also with reference to the cases discussed in the book and, above all, to the experiences of the Sangone RC and the partial projects for the Stura river. The project is a very general framework of reference, extended on a vast scale, durable and suitable for sharing, which is implemented through operational projects hinged to and limited by the strategy. It is a system project which assigns overall sense to specific transformations, taking on a programmatic dimension, projected towards the identification of values for the future and the definition of a vast framework of coherence for the individual projects, implementing transformations using different methods of action and priorities.

In this way, the project becomes a useful tool for orienting development and promoting cooperation, meeting the current needs for selecting actions when there is a lack of resources, acting at various levels and producing an articulate and dynamic design procedure which requires (Regis 2015) a trans-scalar approach to relations with plans and regulations on different scales; multidisciplinary approaches; sustainability, as the governing principle on global and local markets; and the assessment of choices in terms of legitimisation and to encourage implementation (see Brunetta, Chap. 5 in this book).

The river landscape and territorial project are acknowledged for (a) the experimental nature of the methods, which implement many forms of knowledge and technical and/or institutional action, articulated through strategies, orientations, interventions and events linked by a process of participation and social and institutional learning (see Ingaramo and Voghera Chaps. 1 and 3) and (b) a broad system of practices promoted with the involvement of different social actors that the designer or technician interprets with the political part.

This project seems to be open and multidimensional, the outcome of interaction in the Masterplan between different, synergic and sometimes conflicting strategies, which are not necessarily intentional. The project links a chain of actions and interventions in a "decision-making process" in which they interact and express the interests of the local public and private companies and institutions (Albrechts 2004; Healey 2003; Clementi and De Luca 2008). The personal nature of the project, which is simultaneously heuristic, explorative, interpretative and transformative, alludes to the intentional and strategic nature of the project which involves various actors (Pasqui 2005).

From the River Agreement to the enhancement of cultural heritage and landscape in Catalonia (see Sabaté Bel, Chap. 8 in this book), the project is the most effective tool for acting in places where the crisis has hit hardest to grasp the potential of the territory. It enhances the heritage available and promotes long-term sustainable development, to be built up in small parts, laying the foundations for a future that is coherent with the endogenous resources, with the identity of the place and with the expectations of the local subjects. It is able to offer opportunities for qualification, with its very own specifics, within the global system of territories in competition. It is necessary to know how to mobilise socio-economic resources around the shared "vision", directing the single actions towards an organic system and identifying the possible forms of coordination, management and financing of operations.

The examples collected in this book show how a way of planning transformations for the development, quality and safety of the territory emerges, through a series of small-scale independent projects (local projects), developed within the scope of a bigger project (framework project), defined using a strategic apparatus[7] – which is the Action Plan for RCs – and "designed" in a Masterplan which promotes synergies, territorialising them and giving them shared perspectives.

The Masterplan covers territory, creating a network of the existing and possible relationships, proposing organisational and structuring models, taking care of the relationship between architecture and the ground and investing the open collective urban, rural and natural space (Secchi 1986, 1989; Mazza 2010). It promotes project

[7]The River Agreement, applying the indications of the ISPRA National Observatory on river agreements, defines the aims of the design process, beginning with practices: (a) sharing the aims, method and process phases, (b) multidimensional knowledge, (c) definition of a scenario capable of integrating actions over a vast area and also locally and (d) definition of an action plan and schedules (with a timescale of 3 years, to be updated from time to time).

activities that involve the support that sustains the urban settlement, production processes, social action and the deployment of tangible and intangible networks. The Masterplan prioritises the system in a design process which does not establish hierarchies and does not control the tools and methods applied to urban design and architecture. The way of operating refers to a multi-scalar method of "urban design" which takes a fresh look at architectural design within large-scale relational systems (Gasparrini 2013), building up a framework of reference to "planning for projects" (Mazza 1999, 2004; Campos Venuti 1994; Faludi and van der Valk 1994; Gregotti 2000). Urban design implicates a complex, multi-scalar, multifactorial and collaborative design process which interrelates local projects (Healey 1997), developed in relative autonomy, but creating synergies. A broadscale vision flexibly articulates a multitude of small-scale projects, adapting them over time. These projects interact with territorial and landscape planning considered as a frame with variable geometry and dimensions, capable of producing policies and projects to direct change. The effects of the projects depend largely on the capacity of special policies to weave profitable relations with territorial governance tools and ordinary economic management tools, like tax incentives, as well as targeted use of urbanisation charges and innovative use of equalisation systems (Lanzani 2015).

2.3 Landscape and Environment, Territorial Project Themes

Environment and landscape which represent the most stable and durable dimension of the territory are definitely the most desirable outcome, perhaps rhetorically, of the contemporary project. Quality, safety and identity of the places are the aim of the project activity discussed in the book, which promotes and guides processes that transform the spatial, social, economic and cultural context of the territory.

Environment and landscape, successful and locally important narrations of the project, are economic and cultural development factors that bring together care for the territory, cohesion, economic growth and competitiveness through visions which are triggered by the specificity of the contexts, resources (economic and social, environmental and infrastructural, landscape-based and symbolic) and of the territorial actors engaged.

The project activity implemented in the cases discussed in the book – from river agreements to projects for ecological and landscape enhancement of riversides – aims to create new ecosystemic, energetic, food and functional balances, regenerating productive settlements, reassigning a central and multifunctional role to open spaces, restoring the relationships between urban systems and farming and forestry spaces and involving the collective maintenance of the territory (Magnaghi 2005). These are projects which, in keeping with the practice of landscape urbanism (Waldheim 2006), focus on the green and blue networks and on the structuring design of large open spaces (Lanzani 2014).

The project enhances the territorial and landscape identity, redefining the relationships between ecological functions and the aesthetic quality of the territory, and

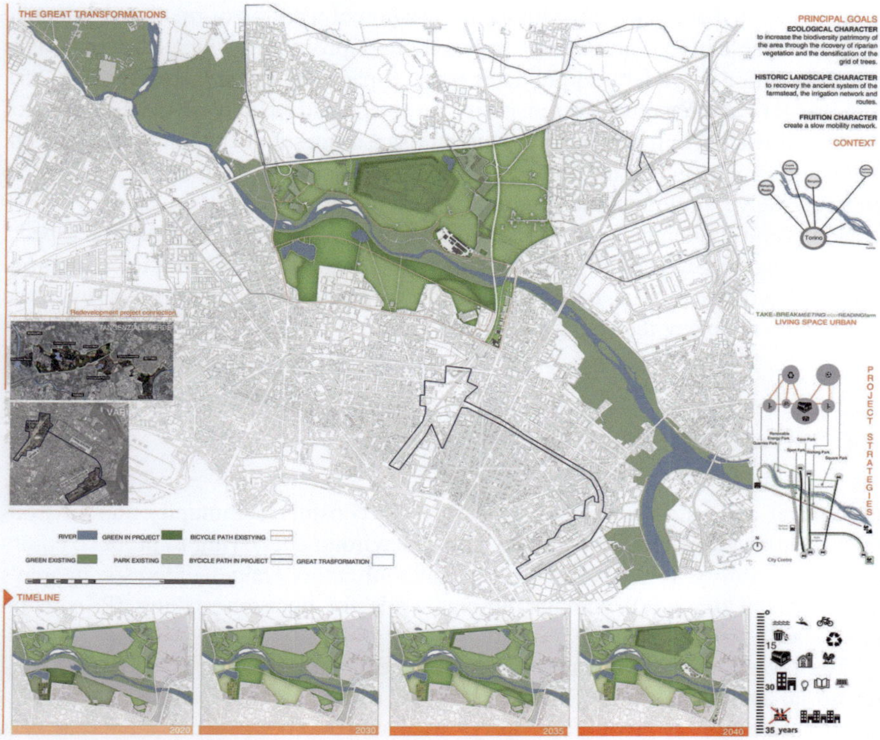

Fig. 2.5 Masterplan of the enhancement of the Stura di Lanzo river, in Turin. Reuse of abandoned or underused areas, at different times (Source: Embrace the Stura experience. International Workshop Design along the river. Turin September–December 2014. Politecnico di Torino. Tutors: Ingaramo R., Voghera A., Claps P., Mutani G.,Mazzotta A. Students: Alfarano E., Borra F., Pignataro A., Mattia Tempesta M.)

acts as a tool for the production of material wealth and social wellbeing, in durable "self-sustainable" forms (Magnaghi and Fanfani 2010). The solutions create a synergy between the urban, rural and natural context, integrating tangible and intangible actions for culture, quality of life, landscape and economy. Multisector, inter-scalar and multifactorial strategies generate development, trying to take the environment and landscape as the basis on which to rebuild the quality of the territory.

Unused or underused resources and spontaneous, resilient and temporary actions play a central role in the territory in times of crisis, plan aside, and they find coherent inspiration in the project.

They use space in brand new ways, confirming the centrality of public space and its quality in the contemporary project (Bianchetti 2007, 2011), a space for inclusion and integration of values and social, ecological and environmental uses. A central role is also taken on by the recycling of structures, tissues and landscapes which have been abandoned and rejected, with the affirmation of a notion of relational, adaptive project, capable of interpreting the cyclic nature of processes, which often envisages the reclamation of underused public or private spaces by the community (Fig. 2.5). The project acquires renewed sensitivity for the landscape and for the ecologies of the territories, to limit the impoverishment of resources that cannot be reproduced, regenerating and regulating the transformations with actions that are resilient to climate change, pollution, consumption of resources and self-organisation.

Social and economic components and environmental, natural and cultural aspects all come together to seek sustainable balances (Dansero 2013) in the inter-sectorial perspective of the project, capable of accompanying the territory, with all its limits and potential, during the change.

2.4 Structural Features of Design

Ecological and landscape planning and design consider community and environmental interactions linking built and natural environments trying to produce equity and sustainability. It needs to consider different actors, values, interests, people's aspirations (CoE 2000) and institutional tools for decision-making (Friedmann et al. 2000).

The project-related visions of the river agreements, *contract de pays* and local territorial agreements don't just rationalise the socio-economic interests of the sector in a projected design of the territory (Magnaghi 2005), like most of the spaceless experiences of strategic planning. They produce processes of change that alter with the alteration of the players involved. The strategies of the Action Plan can correspond to several Masterplans of the groups of interest, and the process of interpretation and synthesis triggered by the various design processes that accompany the evolution of the actions, territorialising them, is absolutely essential (workshop from 2004 to 2010 for the Sangone and Stura di Lanzo from 2005 to 2016).

The projects – including those created by the students on the master's degree course in territorial, environmental landscape planning and architecture involved in the design along the River International Workshop and competition for drawing up the Masterplan and spot actions along the Stura di Lanzo river in 2014 (see Ingaramo and Voghera in this book) – are visions which see the aims of the River Contract and Corona Verde Action Plan differently; behind every design scenario lies the quality of the territory and river landscape in relation to the city and the immediate communication of a "new" urban thought.

Using a linguistic process explained in the Masterplans, the projects express processes of territorial and landscape transformation capable of evolving at different times and in diversified ways, involving the local public and private actors. They are capable of making urbanism clear and accessible and directing it effectively also, thanks to the ability to communicate spatial choices to the public.

This communicative strength of the RC Action Plan Masterplans is inspired by tools that have been developed in numerous countries, like guidelines, abacuses and vast area strategies for territorial projects paying attention to the landscape (Sala et al. 2014), which guide planning on different scales.

In France, the Landscape Charter proposes objectives for actions that can be translated into landscape guidelines and regulations and included in the territorial and urban planning documents, particularly in plans of territorial coherence (Schéma de Cohérence Territorial, SCoT) and in local urban development plans (Plan Local d'Urbanisme, PLU).

In the United Kingdom, Planning Policy Statements and the Local Development Framework (LDF) coordinate how local development is intended to be, including the results of the landscape character assessment (LCA) at local level. Besides green infrastructure (GI), as in the Northamptonshire Environmental Charter and GI Suite and Thurrock Green Grid, the Green Infrastructure Framework Plan is a vast scale landscape starting point for planning and decision-making. It is a multifunctional network of green spaces, both public and private, for improving the quality of life and promoting territorial sustainability. The aim of the GI is to design a strategic network linking green spaces and open spaces with urban areas, new developments and peripheral areas, acting as a boundary for new urban developments while also carrying out a compensatory function.

In the Netherlands, the landscape quality plans (Beeldkwaliteitplan, BKP) – according to Agenda Landschap (2008) – are prepared on the initiative of the public administration, private stakeholders or social organisations; they contain a set of recommendations and guidelines to protect, create and/or improve the landscape quality of a given area, whether regional or local. Interesting examples are Beeldkwaliteitplan Kromme Rijngebied, Beeldkwaliteitplan Sluis-Aardenburg and Beeldkwaliteitplan gemeente Haaren.

The international cases outlined promote landscape transformations using different tools for the strategic communication of the project, which take on the role of integrating the planning tools and increasing social awareness. Interesting from the methodological viewpoint is *Dealing with Change in the Connecticut River Valley* (Yaro 1999), a manual drawn up by the Center for Rural Massachusetts to offer a practical and direct description of strategies for the creative reuse of the territory, with alternatives to traditional planning methods.

The Masterplan in *Dealing with Change* is a rather casual tool, without the restrictions of the plan and the details of the architectural design, which builds a "deal" (Ardielli 2012), a negotiation and an agreement on change, reporting the quality of the physical transformation in a fluctuating and uncertain context. In this case, the project seems to be the means necessary for legible, shareable planning or design capable of explaining the physical transformations behind the strategies. Of course the term Masterplan can refer to the traditional US consideration of urban planning and design (Palazzo 2008), as in Olmsted Jr., who elaborates tools for the design of the water network, the mobility system, which define a different way of designing and living the city, considering the hopes of its inhabitants.

The Masterplan explains the usual spaceless rhetoric of strategic planning, which introduces directions and measures, integrating objectives and actions. It accompanies the construction of negotiations and associations and simplifies competition

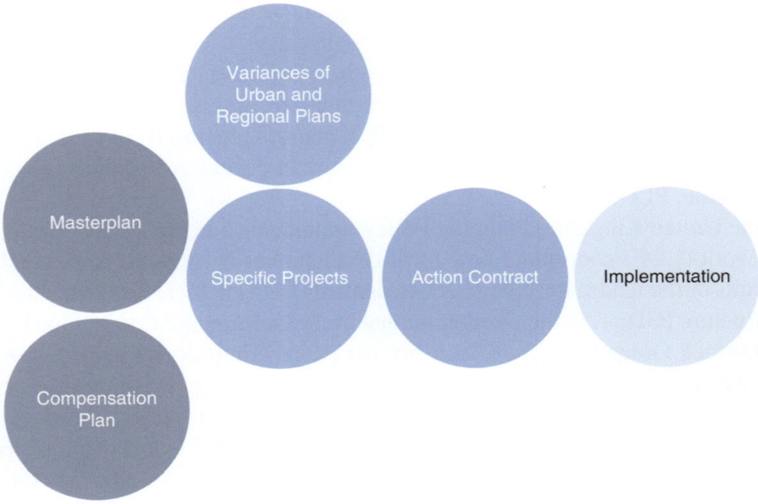

Fig. 2.6 From the masterplan to the action (Source: Author's elaboration)

and the assessment of projects for the selection of alternatives and priorities in implementation.

Strategies, objectives and actions all refer to a design which explains the planning framework of reference and also outlines timing and the rules applied to the selection and implementation of the interventions. In compliance with a broader planning framework, the individual project activities are assessed, implemented and monitored over time in relation to the resources available, which should guarantee their effectiveness and efficiency. With this in mind, the environmental, landscape and economic assessment dimension takes on an important role in explaining the values at stake, relating them to the presumable outcomes of the transformations, contributing, therefore, also to explaining and legitimising the criteria observed in the selection of the interventions (Brunetta and Voghera 2008).

In the case of the RC, it is the SEA, a compulsory procedure, that contributes to the assessment of the possible impacts on the landscape and environment, while the selection of the priorities of the projects is the responsibility of the stakeholders involved in the agreement (whether it is the RC or an agreement for action between public and private actors to implement specific interventions or, where envisaged, as in the case of the Stura river, the Environmental Compensation Plan).

The contractual nature of the projects imposes the construction of special regulatory mechanisms and governance devices which allow the formalisation of associations (Castells 1983; Sanoff 2000) for implementation.

The Masterplan project set-up in this way legitimises values and builds visions of the future (Fig. 2.6), in a framework of inclusiveness and transparency, which simplifies the emersion of alternative processes, arranges "place-based" territorial strategies and allows the social actors to take responsibility for the different objectives and interventions proposed. These are some preconditions to guarantee that the strategic value of the RC can become an operational practice.

2.5 Governance of the Territorial Project

"The territorial project represents subjects, not objects. It unites different points of view, so it doesn't exclude contradictions and conflicts, but places them in the centre. Its strength is the conservation of complexity. Its risk is idle chatter or the legitimisation and hiding of pure strategic action" sustained Giuseppe Dematteis (1995). The multidimensional nature of the conflicts in the dialectics between local and global interests relating to the use of soil (public and private, residential and otherwise) (Mazza 2004), or social interests for the use of space (conflicts relating to forms of spatial and social exclusion and to the supply of services) (Levebvre 1970), as well as those relating to the environment, mobility and defence of identity, is at stake in the project.

The projects discussed in the book develop a complex strategy which involves a considerable number of actors (local and supralocal administrations and institutions and economic and social players), each of which brings their own different vision, interests and resources, by "direct interaction" (Crosta 1995). The project construction approach is "concerted", based on the construction of consent in a plural context (of stakeholders, interests, values and visions of the world) and on the composition of interests and conflicts, as well as the sharing of aims, procedures and implementation processes.

The RC, with its different and articulated projects (Action Plan, Masterplan, Action Contract, Environmental Compensation Plan), is the outcome, and it requires the consideration of the relations between these actors, outlining the process governance methods (Pasqui 2005).

Project governance is the system, which is sometimes shared, as in RCs, to reconcile the interests around which the strategies are built, but it is also the means of inter-institutional coordination, cooperation and partnership between public and private actors who organise themselves into a system of skills, responsibilities and powers. Project governance is transversal to the various project phases: conception, design, financing, construction, management and monitoring.

In the RC, governance is guaranteed by the coordination of the different phases of the decision-making and planning process by the Direction, made up of the most important subjects (usually the institutions) that outline the strategic orientations and guide the implementation process, inviting the Basin Meeting or the Contract Forum, an extended body with which the RC is shared, to participate in specific, crucial moments of decision-making and construction of the implementation process.

The project governance model is nodal, because it promotes the aggregation of the social actors on different territorial scales and defines an open intercommunal status of purpose for the project. The hydrographic basin containing the river and its tributaries brings supralocal players (which, in Italy, are the Regional Administration, Hydrographic District Authority, metropolitan cities, provinces and municipalities) to the geographic scale of the river landscape, in a shared territorial project, promoted by local authorities (metropolitan city and provinces in Piedmont) to link and coordinate the municipalities. The project becomes the privileged place to try and effectively integrate territorial policies through an articulate and complex process of activities that involve a broad range of institutional and noninteractive subjects (Gambino 2010), of diversified resources and procedures that orient the formation of decisions (Bagnasco 2009) and offer a dynamic and plural contribution to the prefiguration of the methods of implementation.

The project – from the Action Plan to the Masterplan and its detailed projects – is a flexible framework (Cremaschi 2003) that houses the relations between different levels of subjects which, depending on the accomplishment of individual interventions throughout the territory, cooperate, going beyond the local government dimension. Institutional cooperation can be organised into different forms of association, agreements and understandings of limited duration, linked to the completion of the project or parts of it, or into stable forms (as in the case of Corona Verde or the River Agreement), which testify to the strategic and aggregative capacity of different interests. Sometimes they are functional aggregations, which can alter the structure of the individual governance responsibilities of the major organisations (Urbani 2008); sometimes they increase the capacity to play and compete in a vast context (like the river agreements, which have been capable, over time, of attracting different regional, national and European resources and cooperating more effectively in networks). The governance process also involves private players and actors from the service sector, who are now essential because they provide self-organised contributions for the management of the landscape (associations relating to community urban allotments, farming associations, environmental landscape defence associations).

Implementation and management are essential and necessarily open components of governance, internal processes of the project which have to change in relation to (sometimes unpredictable) events that redefine the contextual conditions, relationships and strategies of the players (Fig. 2.7).

Management is not, therefore, a technical-executive function of the governance processes and requires piloting methods and mechanisms (like the Direction, the Meeting or the RC Forum as a function of multilevel governance) which oversee the design, implementation and assessment phases; these structures are institutionalised, but flexible and capable of adapting to the changes that emerge during the life cycle of the project, and their job is to direct and to accompany it, to guarantee the progressive and shared specification of the general objectives, guaranteeing the agreements made (e.g. the Landscape Task Force in the Netherlands, which also coordinates the management of the budget and the disbursement of funding)

The multi-scalar, multi-sectorial and multiplayer nature of territorial development projects (Clementi and De Luca 2008) makes the construction of effective coordination structures extremely difficult, while the preventive structuring of a

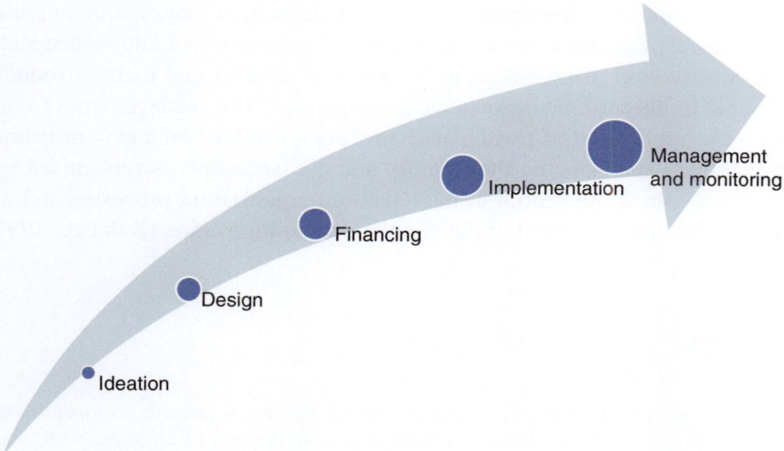

Fig. 2.7 Project governance (Source: Author's elaboration)

multilevel governance system (Hooghe and Marks 2001) makes a positive contribution to the coordination and link between different territorial levels and players.

Governance also has to touch the management of the overall financial architecture (Stanghellini 2002): the integration of different sources, rules and mechanisms of use of funds implicates the capacity to combine actions that depend on different forms of financing.[8] One of the project challenges undoubtedly concerns the ability to produce innovative tools for financing the projects, both in terms of the evolution of project finance tools and from the point of view of experimentation of urbanistic tools for compensation and equalisation (Morano 2010, 2012), the experiments relating to the Stura river Environmental Compensation Plan work in this sense.

The projects linked to RC produce a durable organisational context, based on solid systems of alliances and cooperation among the players concerned or involved, which are also open to allow mobilisation of local and supralocal networks and to activate adequate procedures for the planning and management of operations. The subjects involved (institutional and otherwise) work together towards a joint aim. The consent necessary to the progress of the initiative and the ongoing adjustment of the aims imposes the preventive planning of a context that is favourable to mutual exchanges.

In the development of the project, the result is definitely not acquired ex ante, and, in any case, it has to cope with all kinds of resistance and contestations. Participative practices move in this direction. Making it possible to overcome conflicts in a non-reductionist perspective, they should be considered as open and incremental project exploration strategies, which work effectively on the different forms and

---

[8] Public financial resources (European, national, regional and local resources); private resources (by enterprises, associations and representative organisations, foundations and other social actors); alternative resources (subsidised financing, loans, etc.)

figures of the conflict, offering the project a configuration that is more attentive to requirements. Participation becomes a process of acquisition of knowledge and culture, of construction of relationships between inhabitants and their surroundings, with a view to the informed growth of communities. The management of conflict too, with the application of participative practices, can take on a new instrumental meaning, aimed at enhancing the territory and the landscape, as recognised by the communities, which are central to the territorial organisation processes and a pre-requisite of the empowerment of the population and innovation (Balducci 1991).

References

Affeltranger B, Lasserre F (2003) La gestion par bassin versant: du principe écologique à la contrainte politique – le cas du Mékong, in Vertigo. Rev Sci Environ 14(3), https://vertigo.revues.org/3715 (August 2016)

Agenda Landschap (2008) https://www.rijksoverheid.nl/documenten/rapporten/2008/11/28/agenda-landschap (August 2016)

Albrechts L (2003) Planning and power: towards an emancipatory planning approach. Eviron Plann C Gov Policy 21:905–924

Albrechts L (2004) Strategic (spatial) planning re-examined. Environ Plann B Plann Des 31(5):743–758

Ardielli M (ed) (2012) Masterplan: né piano né progetto. INU Edizioni, Rome

Avidano V, Voghera A (2012) I Contratti di Fiume per lo sviluppo dei territori marginali. Politiche Piemonte 11:12–13

Bagnasco A (2009) La governance dei distretti industriali. In: Dematteis G, Bagnasco A (eds) Le frontiere della geografia. UTET, Turin

Balducci A (1991) Disegnare il futuro. Il Mulino, Bologna

Barca F (2011) Alternative approaches to development policy: intersections and divergences. OECD Reg Outlook 2011:215–225

Bastiani M (ed) (2012) Contratti di fiume. Pianificazione strategica e partecipata dei bacini fluviali. Dario Flaccovio Editore, Palermo

Bianchetti C (2007) Urbanistica e sfera pubblica. Donzelli, Rome

Bianchetti C (2011) Il Novecento è davvero finito. Donzelli, Rome

Bohl C, Plater-Zyberk E (eds) (2006) Special issue: building community across the transect. Places 18(1):4–17

Brunetta G, Voghera A (2008) Evaluating landscape for shared values: tools, principles, methods. Landsc Res 33(1):71–87

Campos Venuti G (1994) La Terza Generazione dell'urbanistica. Franco Angeli, Milan

Carter N (2007) The politics of the environment. Ideas, activism, policy. Cambridge University Press, New York

Castells M (1983) The city and the grassroots: a cross-cultural theory of urban social movements. California University Press, Berkeley

Choay F (2005) L'utopie et le statut anthropologique de l'espace édifié. L'Architecture et l'esprit de l'urbanisme européen. L'Esprit, October:93–113

CIEEM (2016) Guidelines for ecological impact assessment in the UK and Ireland: terrestrial, freshwater and coastal, 2nd edn. Chartered Institute of Ecology and Environmental Management, Winchester

Clementi A, De Luca G (2008) Dal progetto di infrastrutture al progetto di territorio: un'evoluzione necessaria. In: Belli A, De Luca G, Fabbro S, Mesolella A, Ombuen S, Properzi P (eds) Territori regionali e infrastrutture. La possibile alleanza. Franco Angeli, Milan

CoE (2000) European landscape convention. Florence: Council of Europe.
Cowell R (1996) Stretching the limits: environmental compensation, habitat creation and sustainable development. Transac Inst Br Geogr 22(3):292–306
Cowell R (2003) Substitution and scalar politics: negotiating environmental compensation in Cardiff Bay. Geoforum 34:343–358
Cremaschi M (2003) Progetti di sviluppo del territorio. Il Sole 24 Ore, Milan
Crosta PL (1995) La politica del piano. Franco Angeli, Milan
Crosta PL (2010) Pratiche. Il territorio 'è l'uso che se ne fa'. Franco Angeli, Milan
Dansero E (2013) Individui e gruppi: alla ricerca degli attori nello sviluppo territoriale. ESO, Travaux & Documents 35:11–19
Davoudi S, Strange I (eds) (2008) Conceptions of space and place in strategic spatial planning. Routledge, London
Dematteis G (1995) Progetto implicito. Franco Angeli, Milan
Eckerberg K, Joas M (2004) Multi-level environmental governance: a concept under stress? Local Environ Int J Justice Sustain 9(5):405–412
Faludi A, van der Valk A (1994) Rule and order: Dutch planning doctrine in the twentieth century. Kluwer Academic Publishers, Dordrecht
Friedmann J, Chinitz B, Blanco H, Neuman M (1989) Planning, politics, and the environment. J Am Plann Assoc 55(3):334–338
Gambino R (2010) Interpretazione strutturale e progetto di territorio. In: Poli D (ed) Il progetto territorialista. Città, territori e progetti. Rivista del Dipartimento di urbanistica e pianificazione del territorio, Rome
Gasparrini C (2013) Una nuova città: diffusione e densificazione. In: Talia M, Sargolini M (eds) Ri-conoscere e ri-progettare la città contemporanea. Franco Angeli, Milan
Giaccaria P, Rota FS, Salone C (eds) (2013) Praticare la territorialità. Riflessioni sulle politiche per la green economy, l'agroindustria e la cultura in Piemonte. Carrocci Editore, Rome
Gregotti V (2000) Progetto urbano e pianificazione: un dialogo necessario. Urbanistica 104:126–127
Healey P (1997) Collaborative planning: shaping places in fragmented societies. MacMillan Press, Hong Kong
Healey P (2003) Collaborative planning in perspective. Planning Theory 2(2):101–123
Hooghe L, Marks G (2001) Multi-level governance and European integration. Rowman & Littlefield, Lanham
Ingaramo R, Voghera A (2012a) Finding a design method: integrate multiscale project. In: Cities in transformation. Research & design. Ideas, methods, techniques, tools, case studies. EAAE/ARCC international conference on Architectural Research, Milan, 7–10 June 2012, pp 526–529
Ingaramo R, Voghera A (2012b) Planning and architecture. Searching for an approach. Alinea International, Florence
Kuiper G (1997) Compensation of environmental degradation by highways: a Dutch case study. Eur Environ 7(4):118–125
Kunzmann KR (2007) Territoire, Territorialité, Territorialisation. Conference. Institut de Géographie Alpine, Grenoble, 7–8 June 2007
Lanzani A (1991) Il territorio al plurale. Franco Angeli, Milan
Lanzani A (2003) I paesaggi italiani. Maltemi, Rome
Lanzani A (2014) Interpreting regeneration in planning. Urbanistica 154:83–104
Lanzani A (2015) Città territorio urbanistica tra crisi e contrazione. Franco Angeli, Milan
Levebvre H (1970) Il diritto alla città. Marsilio, Padua
Magnaghi A (2005) The urban village. A charter for democracy and local self-sustainable development. Zed Books, London
Magnaghi A (2010) Il progetto locale. Verso la coscienza di luogo. Bollati Boringhieri, Turin
Magnaghi A (2011) Scenari strategici. In: Bastiani M (ed) Contratti di fiume: pianificazione strategica e partecipata dei bacini idrografici. Dario Flaccovio Editore, Palermo
Magnaghi A (2012) Politiche e progetti di territorio per il ripopolamento rurale. In: Bonora P (ed) Visioni e politiche del territorio. Per una nuova alleanza tra urbano e rurale, Quaderni del territorio 2. Università di Bologna, Bologna

Magnaghi A, Fanfani D (eds) (2010) Patto città campagna. Un progetto di bioregione urbana per la Toscana centrale. Alinea, Florence
Mazza L (1999) Trasformazioni del piano. Franco Angeli, Milan
Mazza L (2004) Piano, progetti, strategie. Franco Angeli, Milan
Mazza L (2010) Limiti e capacità della pianificazione dello spazio. Territorio 52:7–24
Minot D (2001) Le projet de territoire. Elaboration et conduite partagées d'un projet de territoire. Bergerie Nationale, Rambouillet
Morano P (2010) La compensazione urbanistica nell'acquisizione consensuale non onerosa di risorse ad uso pubblico. In: Bentiveglia V, Miccoli S (eds) Valutazione e progettazione urbanistica. Metodologia e applicazioni. DEI, Rome, pp 175–206
Morano P (2012) Sostenibilità economica: criteri per favorire la fattibilità degli interventi e attuare la perequazione urbanistica. In: Calace F (ed) Criteri per la progettazione urbanistica. La sostenibilità alla prova. Alinea, Florence, pp 69–92
Palazzo D (2008) Urban Design. Un processo per la progettazione urbana. Mondadori Università, Milan
Pasqui G (2005) Territori: progettare lo sviluppo. Teorie, strumenti, esperienze. Carocci editore, Rome
Pearce J (2003) Social enterprise in anytown. Calouste Gulbenkian Foundation, London
Pettersson H (2004) Compensation within environmental impact assessment in Sweden and the United Kingdom. Cranfield University at Silsoe, MSc in Natural Resource Management, AY 2003/2004, Supervisor: Stephens W
Regis D (2015) Il progetto e la valutazione per lo sviluppo del territorio. Una sperimentazione di metodo in Valle di Susa. PhD thesis. Tutors: Bottero M and Voghera A, Politecnico di Torino
Russo M (2014) Urbanistica per una diversa crescita. Progettare il territorio contemporaneo, Donzelli, Rome
Sala P, Puigbert L, Bretcha G (eds) (2014) Landscape planning at a local level in Europe/La planificació del paisatge en l'àmbit local a Europa. Landscape Observatory of Catalonia. Government of Andorra, Olot
Sanoff H (2000) Community participation methods in design and planning. Wiley, Hoboken
Secchi B (1986) Progetto di suolo. Casabella 520:19–23
Secchi B (1989) Un progetto per l'urbanistica. Einaudi, Turin
Stanghellini S (ed) (2002) La valutazione degli investimenti sul territorio. In: Atti del XXXII Incontro di Studio del Centro Studi di Estimo e di economia Territoriale, Ce.S.E.T., Venice
Urbani P (2008) Governance e pianificazione nella programmazione territoriale d'area vasta: spunti di riflessione. In: Belli A, Mesolella A (eds) Forme Plurime della Pianificazione Regionale. Alinea Editrice, Florence
Voghera A (2015) River agreements in Italy. An experience for river management. recent advances in environmental and Earth sciences and economics. Energy Environ Struct Eng Ser 39:351–362
Voghera A, Regis D (2016) Progetti per territori in trasformazione. Archivio di studi urbani e regionali 177:137–156
Yaro RD (1999) Dealing with change in the Connecticut River valley: a design manual for conservation and development. Lincoln Institute of Land Policy and the Environmental Law Foundation, Cambridge, MA
Waldheim C (2006) The landscape urbanism reader. Princeton Architectural Press, New York

Chapter 3
Urban and Architectural Forms

Roberta Ingaramo

Abstract The contribution is part of the debate on trans-scalar design, outlining a role for design within a system of strategies. The river courses, natural infrastructures yesterday and holders of a denied natural identity today, provide specific scope for reflection, starting from research experiences within the territories of the Metropolitan City of Turin, where design has been accompanied by a River Basin Governance method/tool. Detailed design choices form the backbone for the institutional concerting of actors and specialists in the definition of a system strategy aimed at a new vision of the landscape planning and design process.

Keywords Trans-scalar design • Design along the river • Design process • Relational design • Design strategies • Scenarios • Incremental design • Participatory design

3.1 The Frame

Design research covers themes that are hard to circumscribe to a specific spatial outline or to a single dimension of the matter; the problems take on a trans-scalar order, and the outlining of the scale of intervention is part of the research and of the questions to be answered (Durbiano 2014). Complex problems, like the lack of residential accommodation in the French capital (Atelier International du Grand Paris 2013) and in Bordeaux[1] a couple of years earlier, or urban regenerations like the Euroméditerranée (Epaem 2006) renewal project of Marseille (launched in 1995 and still underway 20 years later) or Battersea Power Station neighbourhood[2] (from 2014) in London, require planning and design capable of adapting to different scales

[1] 5 Dialogues pour 50.000 logements autour des axes de transport collectif. Synthèse de dialogue compétitif, mars 2011. A'urba, agence d'urbanisme Bordeaux métropole Aquitaine. La cub, communauté urbaine de Bordeaux (OMA, Lacaton-Vassal, A. Chemetoff et associés, 51N4E, AUC).

[2] https://www.batterseapowerstation.co.uk/#!/portal

R. Ingaramo (✉)
Department of Architecture and Design, Politecnico di Torino,
Viale Mattioli 39, 10125 Turin, Italy
e-mail: roberta.ingaramo@polito.it

R. Ingaramo, A. Voghera (eds.), *Topics and Methods for Urban and Landscape Design*, Urban and Landscape Perspectives 19,
DOI 10.1007/978-3-319-51535-9_3

in order to meet numerous demands and goals. The difficulty of incorporating several disciplines and targets in a single project introduces a series of communication and acceptance obstacles that must be overcome.

> It is difficult to ensure that what is accomplished at one scale (the design of a discrete building, for example) aligns with values at another scale (such as management and preservation of a watershed). It becomes harder to decide what matters when the priorities set by the project are not native to the various disciplines that are engaged. (Plowright Chap. 13)

All too often, trans-scalar and interdisciplinary matters are rhetorical references to define design strategies, from landscape scale to urban and architectural scale. In most cases, the definition of design strategies on a large scale has nothing to do with urban planning and design on a local city and architectural scale. A delightful master plan or a fascinating strategic scenario gives the illusion of the integration of a system, which will probably be developed in its individual parts through the allocation of building lots to investors, who will commission different architects who will propose their own personal vision of architecture, forgetting the common strategy. An example is the Euroméditerranée renewal project (Dossier Marseille 1994; Samson 1999), where authorial architecture has been used as the expedient of modernity, or Milan's Citylife (the design that won the international tender for Fiera di Milano 2004). Cases in which the design continues through coherently to the architectural farm are very rare. One such case is the acclaimed West8 project (1993–2000) at Amsterdam Borneo Sporenburg, which redesigns two peninsulas in the eastern part of the Amsterdam docks, basing the urban design on spatial architectural choices, which reinterpret the morphology of the historical lot and the typical Dutch canal house.

In this framework, river environments are a privileged area of analysis, in that they are natural infrastructures along or close to which the cities have been built up and which, in recent years, have become major subjects of regeneration processes. Along the big rivers, in some cities, projects capable of triggering long-term alteration have been developed, such as the Three Rivers Park [3] in Pittsburgh which, in 1999, began radically changing the relationship between the city's buildings and the natural infrastructure, defining a new morphology of the urban riverscape (Carter 2015). The design of the *Parque Lineal del Rio Manzanares*, the urban stretch of which runs through Madrid[3], following the underground part of the M30, has enabled the redefinition of spaces freed up from infrastructure which are part of a bigger urban design focused on the river (Burgos et al. 2014).

A more recent case is the Minneapolis Riverfront design competition (2011)[4] where the recovery of a long river characterised by extensive abandoned industrial settlements required a preliminary design for a transformation which related environmental ecological aspects to the development of the city, which reclaimed a relationship with natural infrastructure. The designers' considerations, "There's

[3] Subject of an international tender in 2005 assigned to MRIO Arquitectos Asociados S.L. (Burgo & Garrido, Porras La Casta, Rubio &Alvares-Sala) with WEST8.

[4] Won by TLS/KVA RIVERFIRST, http://minneapolisriverfrontdesigncompetition.com/

garbage. It's a lonely place, what we weren't prepared for was the degree of separation. The river has been more like a truck dock." (Mack 2014), highlight the fact that the river cannot be recognised as a natural element, considered as a place of rejection and estrangement.

Some of these projects reveal a detachment between prefiguration and design, as in the case of the River City Gothenburg Vision[5], for Gothenburg in Sweden, which focuses the regeneration of the entire port city on the river and the relationship with the waters, the gradual rising of which is also recorded in the new building regulations[6]. It proposes strategies which, while wanting to be highly innovative, have given first results which continue to be far-removed from expectations, redesigning the city with repetitive blocks that have no relationship with the water and with a system of public spaces concentrated at the end of the piers (Cuff, Dahl 2015) . A different approach is taken by the *Plan des transformations de l'Île de Nantes*, where a transformation project that tackles the theme of extensive reuse of an abandoned industrial naval area was initially activated with the recovery of certain buildings (Chemetoff and Berthomieu 1999). It then placed the Loire at the centre of a complex and trans-scalar territorial, urban and architectural project, led by Alexandre Chemetoff (until 2010), which is currently in its second phase (2010–2020 Phase 2[7]) (Chasseriau 2008).

The Seine too is now the focus of a series of ideas, *Reinventer la Seine* (scheduled for completion in autumn 2017)[8], which, thanks to a strategy divided into parts, is seeking proposals for 41 sites scattered along its length, with very different characteristics: from the ports to the riverbanks, from abandoned industrial sites to water treatment installations, and from ecological districts to an abandoned winery. The river is an opportunity to rethink different themes which all have to do with the theme of large-scale natural infrastructure.

After using the water courses as an industrial dumping ground (see the American cases already mentioned or European cases such as Emscher and, to an even greater extent, those in oriental countries like China or Korea) or for intensive agricultural exploitation for about a century, the environmental question has become an essential guide to the development of our territories.

The transformation of Emscher Park in the Ruhr[9] and of the river's tributaries is a paradigm of reference at international level and, after more than 20 years, is still continuing, expanding its range of action. It's an in-depth and timely regeneration

---

[5] Combination of actions and strategies begun in 2010, to guide the transformation of the areas along the riverside, once populated by shipyards and industries which are now largely abandoned. The strategy envisages the regeneration of the quarters of Lindholmen, Södra Älvstranden, Kvillebäcken, Backaplan, Gullbergsvass, Ringön and Frihamnen.

[6] The accesses to the new buildings have to be installed +2.8 m above the average water level.

[7] http://www.iledenantes.com/en/

[8] http://www.reinventerlaseine.fr/en/

[9] Guided by IBA Emscher ParK L.t.d., a consulting company active since 1989, shut down in 1999, the various regeneration operations in the region were led as part of a participated project with institutions, social groups and business groups in the area.

of the German region, now guided by a strategy, Concept Ruhr, elaborated since 2007 (IUAV journal no. 134, 2013) and involving three counties and 35 cities, in a vision of transformation ongoing until 2020. Adaptive reuse for industrial buildings and art in the landscape has contributed to Emscher Park's success.

Not just large watercourses are involved though. Steams and canals also become a privileged area of intervention, sometimes to restore a natural appearance to severely congested cities. One such city is Seoul, where, in the early years of the twenty-first century (2002–2005), the mayor and future president of Korea, Lee Myung-bak, promoted the recovery of an old stream, the *Cheonggyecheon*, which flows east to west in the heart of city, eliminating the infrastructure that had covered it and giving the city back a new area of public space structured into a linear park. The benefits of the operation have been huge in environmental and social terms, and also with regard to the value of real estate, and about 53,000 people use the park every day.

Similar operations have been carried out on the Zhangjiagang in Suzhou, Jiangsu in China, where the river had been overwhelmed by urban expansion, and in our Dora Park in Turin, where part of the Dora river that had been covered over in the 1950s has been uncovered. In Marseille too, the design by Leclercq (Euromediterranee 2, L'extension, F. Leclercq 2011[10]) for the new ZAC, uncovered the Ruisseau des Aygalades and designed a park which has become a detainer basin in the event of flooding, creating a design which follows the altimetry of the water table (Bertoncello and Dubois 2009; Molga et al. 2014). The ground design (*disegno di suolo* Secchi 1986) seconds the uncertain flow capacity of the waters and the physical progress of the torrential stream, which becomes the privileged eco-sustainable view from the new residential settlements of the ZAC, overlooking the new environment which has almost nothing natural about it but does a very good imitation.

Italy's universities have also been called upon to participate in research projects for large-scale urban riverside regeneration, like in the case of Zhongshan in China, where the uncovering of an ancient river in the city centre, the Jiu Qu River, takes on the form of a large-scale urban project[11], which follows in the wake of the success of a big park project, the Zhongshan Shipyard Park (11 hectares, 2002), designed by the Turenscape studio, on the site of a large, abandoned shipyard.

Over the past few decades, rivers, streams and canals have gone from being natural infrastructures, resources to be exploited, initially for transport and the procurement of water and then for the dumping of waste waters from urban settlements, farming and industrial settlements, to become an important theme of reflection

[10] 2009 Competition, 2010 Plan guide, 2011 Prima ZAC, nine-year framework agreement, Marseille.

[11] On which several Italian universities have worked (Italian Schools of Architecture (ISA), Faculty of Architecture of Pescara, Università d'Annunzio of Chieti-Pescara, Faculty of Architecture of Ascoli Piceno, University of Camerino Faculty of Architecture at the IUAV in Venice and the Local Planning Institute (Zhongshan Urban Planning & Design Institute, ZSPDI). The work was published by Clementi et al. 2014.

for design which combines the ecological-naturalistic and urban-architectural dimension, as a tool within the decision-making process for selecting elements and values (Plowright 2014) and to develop and express community visions for a new living space and environment. The spaces we live in are full of nature, not only in dedicated, protected parks and gardens but endemically and virally on walls and roofs and on abandoned infrastructures; water is a precious element which becomes a means for reclaiming, irrigating, restoring and rebuilding composite landscapes.

3.2 Design Experiences for Composite Territories

3.2.1 Telling the Story of Spaces

From 2009, my research initially concerned the Sangone, a torrential stream that runs through the southern outskirts of Turin. Then, in 2014, it moved to include the Stura, another torrential stream which touches a larger part of the city, in the northern area, partly included in the plans of Variant 200 to the local planning scheme, which is very fragmented and associated with complex functional differentiation.

Turin, as stated in the title of one of the projects promoted by the Municipality of Turin for the restoration of the river banks of its four rivers, is a "water town"[12]. It is geographically located in an area crossed by four rivers, two of which, the Sangone and the Stura, relate only partly to the weakest fringes of the city which began to approach their banks only from the mid-nineteenth century (Fig. 3.1).

At the beginning of the twentieth century, small bathing facilities were located south of Turin, at Mirafiori, along the Sangone river, and in the northern part of the city, the sandy banks of the Stura river filled up on summer weekends with those who couldn't afford to go to the seaside. The bathing facilities along the Po river, towards Moncalieri, resisted until the 1960s, when, with the economic boom and the construction of the Italia 1961 district, to commemorate the 100th anniversary of the unification of Italy, the people of Turin began travelling to the beaches of Liguria. The landscape in the river basin which, in the eighteenth and in the early nineteenth century, was characterised by the presence of farmhouses and agricultural plots of a certain importance, retaining a natural appearance, was transformed

[12] Città di Torino, Assessorato alle Politiche per la casa ed il Verde, Settore Grandi Opere del Verde Pubblico. *Torino Città d'Acque* is the project approved in 1993 by the Municipality of Turin which envisages the recovery of the riverbanks in a single river park running for a length of 70 km, with a surface area of 17 million square metres. The operation connects Turin's four rivers (Po, Dora Riparia, Stura and Sangone) to create a continuous system of river parks connected by networks of pedestrian and cycle paths, nature and educational trails, with the defense and enhancement, for each river, of the environmental and architectural details. Opening up subsequently to the territorial scale, the plan envisages the creation of a transition area between the parks in the city centre and the larger parks on the hillside and in the outskirts, as far as the regional parks in the suburban area, connected via the "Corona Verde" with the farming and forestry context of Piedmont's valleys.

Fig. 3.1 Turin and its rivers: north the Stura river as an urban limit and more in the town the Dora river; east the Po river and the hills; south the Sangone river (Source: author's elaboration)

from the 1950s by the installation of industrial and commercial settlements along the access roads, compromising its environmental layout. Highly pollutant industrial complexes (OMA and Industrial Chemicals along the Sangone river) and the AMIAT solid waste dump north of the Stura river, considerably changed a natural layout which had stayed the same for centuries, so much so as to prevent bathing and fishing (Sangone).

The Sangone and Stura are natural infrastructures that determine territories geo-morphologically characterised by alpine and hilly environments, substantially different from one another, with a vocation for tourism at higher altitudes while being severely compromised on the plane, where there are more extensive irrigation systems, the major towns have developed and there are numerous industrial settlements. Many of these have now been abandoned, having changed the ecological balance of the watercourses and land in a very short space of time (from the 1950s).

The Sangone, which stretches for a length of 47 km from the Sangone Valley to the Po river in the southern part of Turin, in the Mirafiori district, home to the mausoleum of *La Bela Rosin*[13], runs through a composite landscape which is much more urbanised in the final flat stretch. Along its course there are farmhouses, Drosso Castle, farmland, industrial estates (in Coazze, Sangano, Rivalta, Beinasco, Orbassano, Nichelino and Turin), the San Luigi hospital complex and the Sito logistic centre. The *Miraorti* project was partly completed along its last stretch in Turin (2010). This is the product of research accompanied by a participated design procedure in support of the administrations[14] with a transformation scenario that envisages an Agricultural Park in the area between Drosso Castle and the Mausoleum of *La Bella Rosin*, supported with the creation of urban vegetable gardens, which are now regulated, in place of those abusively created in *Strada del Drosso* and near the Mausoleum, as well as the transformation of Piedmont Park and its Farmhouse into an agricultural park.

The territory along the Stura river, which stretches for about 69 km when the torrential stream is at its fullest capacity, to its confluence with the Po river, has a varied landscape which extends from the Lanzo Valley to La Mandria Park and Venaria Reale with its *Reggia* (UNESCO site since 1997) and reaches the city of Turin in its final stretch. Unauthorised urban allotments, gypsy camps and industrial settlements are situated on its banks, crossed by road and railway bridges which overlook a bed made up of small gravel and stone islands, as far as the confluence with the Po river on the border with Settimo Torinese. The south bank of the Stura in Turin forms a natural boundary with the residential constructions, merged with commercial buildings and warehouses, and manufacturing units, which are much more intensely located on the north bank (Iveco, Michelin, Pirelli Grandi Turbine Torino, Auchan, Panorama, TNT Traco). The structure of the tissue characterised in this way by important industrial plants is altered even more by the presence of monofunctional satellites, like the Falchera district[15] or the Basse di Stura dump, which create a complexity in which farmland provides a backdrop, as the only link connecting realities that are distant and profoundly diversified. The urbanisation that touches the south bank of the Stura river is the fragile boundary of the city of Turin, characterised by a variety of functional (residential, mass retail, recreational, productive and storage) and social constructions which determine a highly composite urban form, the product of planning rationales which adapt poorly to the need for flexible processes that are subject to time limits.

Environmental system, urbanised system and infrastructural system morphologically characterise the basins, in cohabitation with balance-related difficulties that are clearly visible in the description of these places. In the recent history of urban development, the territories overlooking Sangone and Stura have been treated not as

[13] Built in 1888 as the grave of Rosa Vercellana, morganatic wide of the Italian King Vittorio Emanuele II of Savoy.

[14] http://miraorti.com/

[15] Residential district built by INA casa to a design by Giovanni Astengo (Astengo, Molli Boffa, Passanti, Renacco, Rizzotti 1952–54, with construction of the central nucleus in 1959), and then expanded, becoming Falchera Nuova in the 1970s, designed by Rizzotti, Bianco, Nicola and Romano.

viewpoints and environmental lungs of the city but as backyards or resources to be exploited. The floods that have caused the rivers to break their banks on several occasions, in 1962, with the collapse of the bridges at Mirafiori and Stupinigi over the Sangone, and later in 1994 and 2000 over both rivers, have drawn attention to the need to rethink these landscapes and their relationship with the buildings in the area.

The Planning Scheme of the City of Turin (1995) envisages restoration and transformation of the industrial areas of Basse di Stura and the Dump into a park. An inter-municipal park is also planned by the municipalities of Settimo, Borgaro and Turin, and this will be known as the Tangenziale Verde (PRUSST), a sort of greenbelt prioritising environmental transformations, which are currently hard to accomplish, being far from feasible at a time when resources (especially public) are limited. But the connotation of the flat stretches of these river areas is primarily urban, with a *territory subject to dispute between forms of sharing tied to legal and illegal and evident and hidden practices* (Bianchetti 2012); private, institutional and unauthorised farming areas; manufacturing activities, many of which are being increasingly abandoned; and gypsy camps.

The large-scale territorial projects (Piedmont Region's *Corona Verde*, *Torino Città d'Acque* promoted by the Municipality of Turin, the *Tangenziale Verde* Project managed by the PRUSST of Settimo, Borgaro and Turin, and the P.E.R.A) which have involved the urban belt along the Stura and part of the Sangone, are mainly focused on themes of the urban river park or metropolitan ecological systems, shifting the vision of system logics. These projects take us from a construction of the built-up environment in successions of blocks, solids, areas with single functions linked by an infrastructural system focused exclusively on mobility and accessibility, to an interest for the connection network, greenways and potential opportunities for the requalification and definition of locations. (De Rossi and Durbiano 2006).

These strategies require public resources which are currently hard to obtain. With all of this in mind, it seems more feasible to shift attention to themes and projects that make it possible to solve problems and contingent needs, sometime of a temporary nature, but which can promote change, triggering processes of redefinition at urban and social level that are more incisive and immediate, with short or medium-term effects. The need to integrate the naturalistic ecological project with urban design (Grahame Shane 2011) and architecture now leaves space for scenarios that are transversal to the river course, where nature is the structuring element that defines the river park, becoming a detention basin to convey physical consistency and weight to certain parts of it, like in Rivalta, Nichelino and the Mirafiori area of Turin along the Sangone river or the Turin area of Basse di Stura or the Arrivore Park along the Stura river.

3.2.2 From a Project in Parts to a System Vision

The international panorama of designs along the rivers accomplished or undergoing definition highlights the need to create project scenarios which are part of a broader strategy but which have their own independence in terms of implementation and a controllable spatial dimension.

The decision to start from a project in parts is based on the desire to take into consideration local designs which have implemented the expectations, desires and needs of stakeholders, in a process which starts by assuming the outcomes of the participative processes of the Sangone River Contract (2008–2009) and the Stura River Contract (2014–2015) method.

In our research, the identification of certain Strategic Contexts[16] is just an apparent alteration of the structure of the system in order to further design scenarios which fall into distinct geographic areas (Dematteis and Governa 2005) and which forget about institutional and administrative boundaries, which do not adhere to the conformation of the landscape, focusing on physical, functional, economic and cultural characteristics.

Several case studies have been identified along the Sangone, in the Metropolitan area, such as the severely polluted area of Rivalta or the outskirts of the built-up area of Orbassano, forgetting about the stretch within Turin, subject to some recent transformation processes which have particularly concerned the Sangone Agricultural Park (see Miraorti).

In the case of the Stura, research is focused, instead, on stretches of the river which have been altered more intensely by man, in the areas which go from Venaria Reale as far as the confluence with the River Po in San Mauro Torinese, with a long stretch through the northern part of the city of Turin. This territory is located between two large areas of transformation, *Variante 200*, a modification to the Planning Scheme in force[17] and the area occupied by the Michelin plants along Corso Romania[18], on the administrative outskirts of the city, towards Settimo Torinese. Both projects are still on paper, and the city's altered political situation[19] with a low-cost growth model, accompanied by consistent difficulties in finding

---

[16] With the contribution of the River Contract Action Plan (see Chap. 2 Voghera).

[17] *Variante 200* (2010) occupies 900,000 square metres of space to be requalified in the city's northern quadrant. The former "Scalo Vanchiglia", the railway trench which extends as far as the "Parco Nord" rail junction and the Rebaudengo "node", with the underground station and the square, the urban entrance for those arriving from Milan. The Masterplan of the Variant was drawn up by the design team to Make! Led by the Turin-based company RecchiEngineering, winner of an international tender (2012). The team is made up of Architekten Cie, Tra and Mesa.

[18] The Michelin and Cebrosa areas, totalling approximately 600,000 square metres, located north of the Stura, along Corso Romania, envisage the transformation of an area with a mainly manufacturing vocation, thanks to the adoption of two Integrated Programmes (PR.IN.), into a complex with company management offices, residential accommodation, some social housing, service units and services for sport and leisure, all incorporated into a network of public spaces and pathways.

[19] In June 2016, Chiara Appendino, a member of the Movimento 5 stelle political party, was elected mayor of Turin and of the Metropolitan City.

investors, does not make them feasible in the short term. The relationship with areas of transformation of such extent suggests the space between the formulation of strategies which, remembering the context, can offer possibilities of independent operation capable of triggering regeneration processes from the inside. These can also be self-triggered, without imposing extensively and expensive transformations, but only those of a partial nature, which can be implemented over time, allowing for the alteration of the aims during their accomplishment by phases.

The space in between is a place for possible experimentation, paying particular attention to that third place, which is neither public nor private, that no-man's-land, which seems to remind us, in terms of definition and action, of the third landscape of Gilles Clément (Clément 2005).

3.2.3 Pervasions: Living Landscapes

The image of the river as an ecological infrastructure, a river park capable of restoring a natural identity, rebuilding a landscape, clings onto a system vision, in which everything is park. But the design of a park which follows the course of a whole river seems to clash with the huge diversity of the places, as well as the difficulty in finding adequate resources for planning which is reduced, in the best case scenario, to the construction of cycle paths and footpaths and to works for the re-naturalisation of the riverbanks. The incredible success of Emscher Park is partly due to the location of effective and timely operations, capable of transforming a narrow, controllable panorama, at least initially, and becoming endemic as time goes by, and each of them gains recognition. The landscape cannot be grasped as a whole, offering only partial views (Corboz 1985), and it is this breakdown that creates projects in parts, the design of which is not the result of a zenithal vision, being three-dimensional and adaptable, capable of changing as time goes by.

This is the case of the park along the Ceronda river, which joins the Stura at Venaria Reale, a residual space that has now been transformed into a large parking area at the service of the Reggia; like an overflow area it expands, finding its way in the historical urban tissue, where it becomes part of the reuse of abandoned buildings, building up a system of relations which reveals the presence of the park in the consolidated city and beyond (Fig. 3.2).

The strategy envisages the construction of a park designed to be the continuation of the urban tissue, along the Ceronda river, like in the previous stretch towards Turin, where the buildings form a curtain. Planted boundaries indicate the presence of different materials that alternate in the construction of a system that clings to buildings and ordinary roads and paths, redesigning the spaces and the structure (Figs. 3.3, 3.4 and 3.5).

Filaments of a linear park along the watercourse insinuate and build a change, a reuse of underused buildings, a design of the public space (Piazza Don Alberione) with its road system. Permeable materials are incorporated into the built-up area with a design in which nature becomes architecture (Fig. 3.6).

Fig. 3.2 Venaria Reale (Metropolitan City of Turin) pervasion strategy: the park is in the town (Source: author's elaboration)

Often, what would like to be a river park is not recognised as such. Eighty one percent of the Corona Verde funding has been allocated to improving the system of fruition that has unfortunately been transformed into the construction of cycle paths, many of which have been reabsorbed by the vegetation within the space of just a few years, which does not recognise them as a permanent structural part of a system. An example is the Navarra Linear Park (1999 Caltagirone/San Michele di Ganzaria CT). A river park is not just a system of cycle paths and footpaths. It is a design which allows recognition and specific characterisation depending on the places, with rows of trees that grow up and thicken to offer scenarios and borders to a public space that pushes from the city towards the riverbanks (Bianchetti 2007), with views defined by a nature that guides, like in Parc aux Angeliques, in Bordeaux.

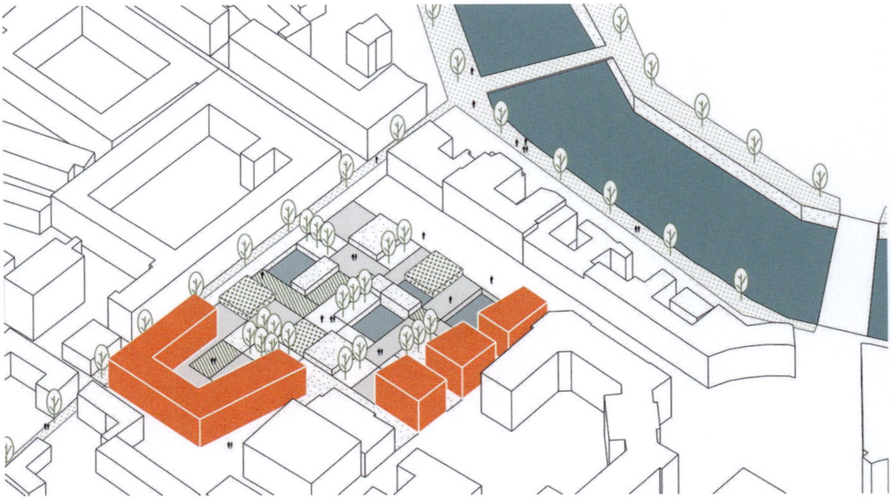

Fig. 3.3 The strategy redefines with different materials the public space (Source: author's elaboration)

Fig. 3.4 From the city to the river banks, the ground design with natural elements (Source: author's elaboration)

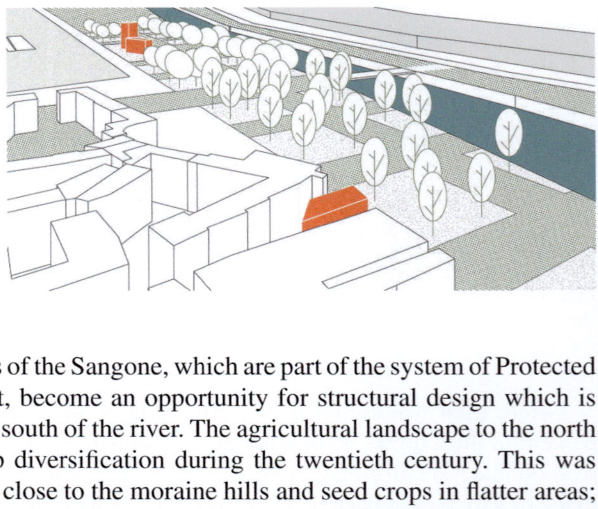

In Rivalta, the two banks of the Sangone, which are part of the system of Protected Areas of the Po River Belt, become an opportunity for structural design which is diversified to the north and south of the river. The agricultural landscape to the north gradually lost its rich crop diversification during the twentieth century. This was characterised by vineyards close to the moraine hills and seed crops in flatter areas; some of the farms, which were built to a plan featuring an inner courtyard, a typical expression of the farming system managed by Piedmonts aristocracy that sprang up as of the eighteenth century, have been abandoned. One of these, Rifoglietto, protected by the superintendence, has been bought by the Municipal Administration.

South of the Sangone river, the localisation of severely pollutant industries, OMA and Industrial Chemicals, for the treatment of oils and solvents, compromised the ecological balance of the soil, water and air during the last 40 years of the twentieth century, with thermo-destruction systems that have now been restored. The

Fig. 3.5 Into the city nature become architecture (Source: author's elaboration)

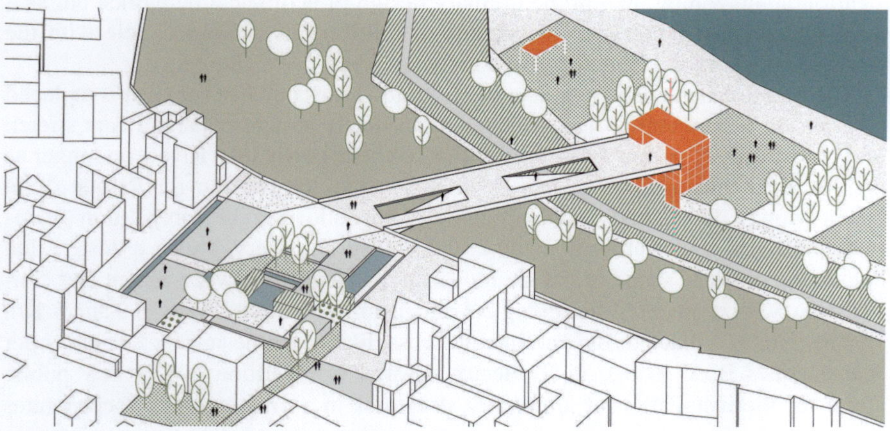

Fig. 3.6 Different levels are linked by a system of public space in the fringe of the city, on the river banks following the natural geographies. Light structures that overcome differences in height (Source: author's elaboration)

fragmentation is emphasized by the presence of a small industrial area and an animal shelter, next to a large farming area.

The project, which has assumed a vision of transformation of the territorial and landscape system of Rivalta, at different times compatible with an architectural and landscape-natural transformation (2016, 2025, 2050), has envisaged a dual strategy with a different characterisation of the two banks. The recovery of the ecological dimension is implemented via the reconstruction and strengthening of the design of the agricultural tissue, in an Agricultural Park, historically shaped by man north of the river, and the design of a contemporary eco-park south of the river where the abandon finds a project of a weak urbanisation, an echo of the design by Andrea

Branzi[20] (abandoned area belonging to Philips in Eindhoven 2000, repeated in the Tender for the reorganisation of the gardens at Porta Nuova in Milan 2004)[21] with an adaptive reuse of the buildings that fit into a design in which the cultivations for the production of energy (miscanthus, sunflowers, etc.) generate a new ground morphology. The endemic crop landscape and the system of historical farmhouses in the Agricultural Park relate with the signs of industrial memories, via a network of irrigation channels. A "fan" designed and hinged on a Silos, now a vegetable greenhouse, assigned the status of landmark of communication of the past and the innovation of Rivalta, stemming from the production of renewable energy, the diversification of economies, functions and images. The new dimension of the landscape has to be perceived through a design of channels at different heights, which reach the river via jetties projected onto the Sangone. The shells of the industrial buildings redesigned as a centre for youth, social, fruition and productive activities linked to innovation. The design of the park, a double detention basin of a reified naturalness, capable of physically structuring two landscapes with different physical features, is not seen as a cartographic zoning, like in the ordinary design of ecological networks, but as a three-dimensional representation capable of fulfilling its ecological role with the recovery of a landscapist perceptive dimension (Figs. 3.7, 3.7a, 3.8).

The traditional vision of a project which follows the course of the river is replaced by a planar, transversal project which can physically structure a significant stretch of river territory to regain a naturalness that has been partly lost. Rivers no longer as merely linear infrastructures, like Toyo Ito, in the competition for Parque de la Gavia (won in 2003), had been pointed out – not following the competition guidelines but as an opportunity to rethink and, where necessary, create a design that could take the very shape of the land to sustain a balance between nature and construction, with respect for the ecological dimension (Ota 2004) (Figs. 3.9 and 3.10).

The river is a place to live, not just to cross, like in the design by Khoolhas per Jean-Jacques Bosc Bridge in Bordeaux (Design Competition 2013), new public space for the installation of temporary structures in a project which incorporates water into the design of the places, respecting their times and alteration and improving the possibilities for structuring.

We should no longer consider banks only to contain, but fringes that offer possibilities for a widespread renaturalisation, capable of welcoming extraordinary natural events into detention basins, lamination belts that redefine more flexible borders and redesign varied places, capable of holding under ordinary circumstances, also with light structures that overcome differences in height, but which can be occasionally reabsorbed by the watercourse. These are places from which to start

[20] With Ernesto Bertolini and Lapo Lani, Design Favela High Tech, Masterplan for Eindhoven, a metadesign for the areas abandoned by Philips, with a weak urbanisation model made up of a productive agricultural territory that could be crossed with research laboratories, residential accommodation, leisure structures, an enzymatic territory with no stable image that changes its layout with the seasons, the way the agriculture around it does. http://www.andreabranzi.it/portfolio/favela-high-tech-masterplan-per-eindhoven-andrea-branzi/

[21] Italo Rota and Andrea Branzi (with Ronan and Erwan Bouroullec) widespread weak modernity.

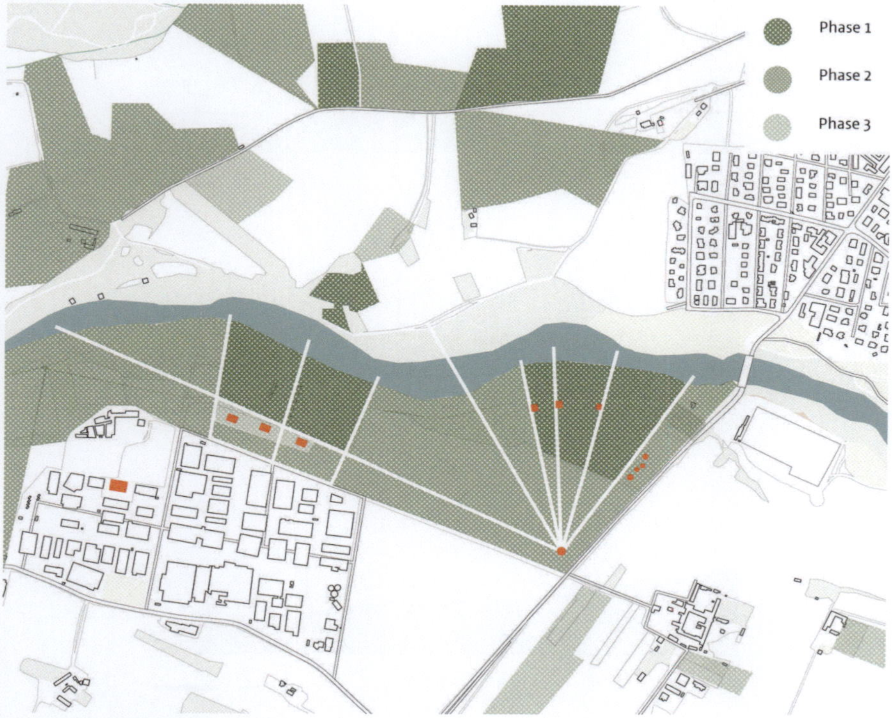

Fig. 3.7 Rivalta (Metropolitan City of Turin), the two river banks, incremental project to redefine landscape and architecture, a dual strategy (Source: author's elaboration)

to redefine the relationship with built elements, which should never be seen as foreign elements, but as something relational, capable of adding value.

3.2.4 Limits and Permeability

The theme of the limit/border takes on a strongly tangible importance in rivers; natural infrastructure is a barrier to buildings in the urban context, but depending on the type of river and the type of course, a different spatial relationship is created with urbanity. In the case of the Sangone and Stura, the torrential nature of their courses, with differences in altimetry that are sometimes considerable (in the stretch between Venaria and Turin, for instance), has influenced the determination of a distance between the riverbed and the first buildings. As I have already pointed out, they have not been considered as a natural place to relate to but as a backroom, where "secondary activities" are located, from shooting ranges to vehicle depots and from goods storage to wholesalers of materials and animal shelters. These places have been occupied by illegal practices (gypsy camps and unauthorised allotments), which

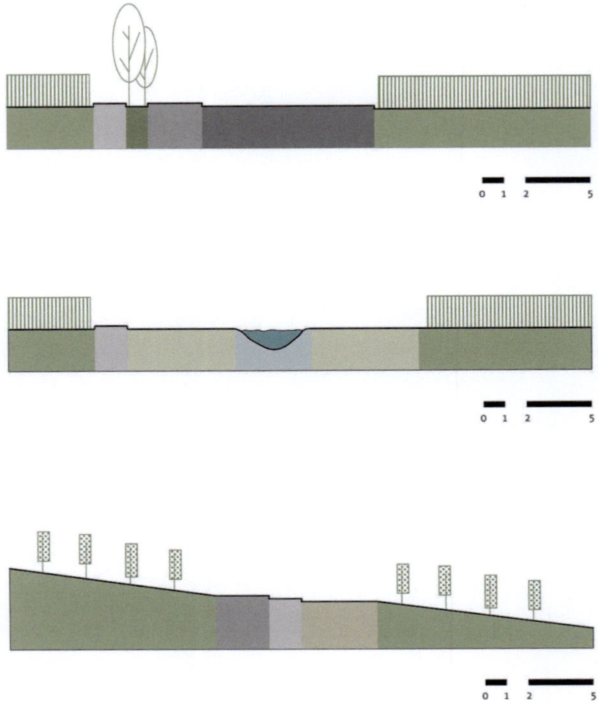

Fig. 3.7a Rivalta (Metropolitan City of Turin), different paths from the top: along the existing infrastructures, agriculture, pedestrian path, green frame, bicycle path and infrastructure; along the canal, agriculture, pedestrian path, flowery meadow and canal; in the vineyards, vineyard, pedestrian path, bicycle path, dirt road and vineyards (Source: author's elaboration)

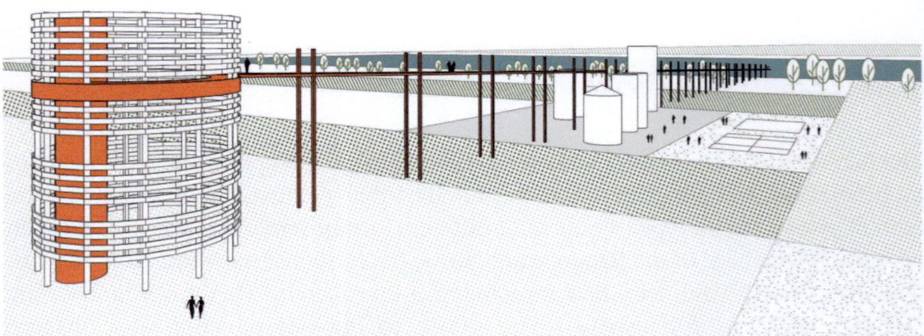

Fig. 3.8 The Silos, looking to the river and the hills on the other bank, landmark of communication of the past and the innovation of Rivalta (Metropolitan City of Turin) (Source: author's elaboration)

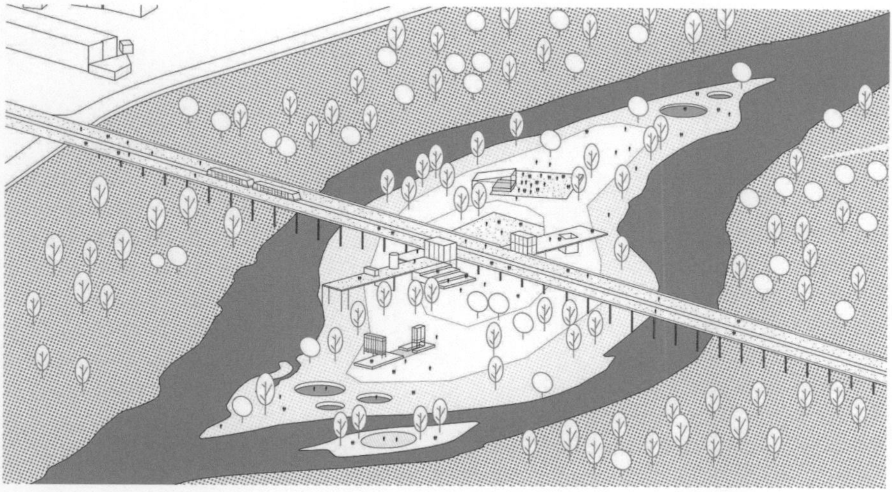

Fig. 3.9 A bridge is a place to pass by but also to stop and stay, a new public space, crossing even small islands that can be used for temporary events (Source: author's elaboration)

have accentuated their breakdown. The *Corona Verde* project[22] affirms the importance of the urban fringes, also in environmental terms, but forgets that they exist when it comes to financing local projects, forgetting that an environmental system is structured starting with the edges and with the design that determines them.

But the urban fringes that relate to the river are not presented as a defined and rigid curtain, as a threshold to defend or to cross. They don't want to become a barrier around urbanisation, as theorised in a now obsolete project by Francesco Venezia for the Lauro plan to "wall in the town, while building a bridge between the town and the rural nuclei"[23]. Regardless of the formal outcome, Venezia's attempt at starting from the architectural form, designing buildings and their spatial positions and then determining the relative urban planning parameters to prefigure a scenario of urban-architectural transformation through ordinary planning, was interesting. The concept of the limit is overcome in our research by a strategy which sees architecture as a starting point for the design of this intermediate space, threshold between the re-naturalised riverbank and urban fringes with no specific architectural characterisation (Fig. 3.11).

The scenario for an urban boundary of Orbassano and Rivalta is an opportunity to identify a position for a project which places nature and architecture on the same level,

[22] Strategic regional project launched in 2003, the Direction Scheme was drawn up by Turin Polytechnic (2007) with the aim of creating a green infrastructure which incorporates the "Crown of Delights" of the Royal residences into the greenbelt, represented by the natural heritage of the metropolitan parks, the rivers and the rural areas which have changed very little, to requalify the metropolitan territory of Turin and improve the quality of life.

[23] The appointment for the Regulatory Plan for Lauro, a municipality in the province of Avellino, with 3700 inhabitants, is dated 1981. The work team was made up of Antonio De Pascale and Francesco Venezia, and Venezia's studies, which analyse the theme of am "untidy" periphery which invades a farming landscape, are very interesting (Venezia 1984).

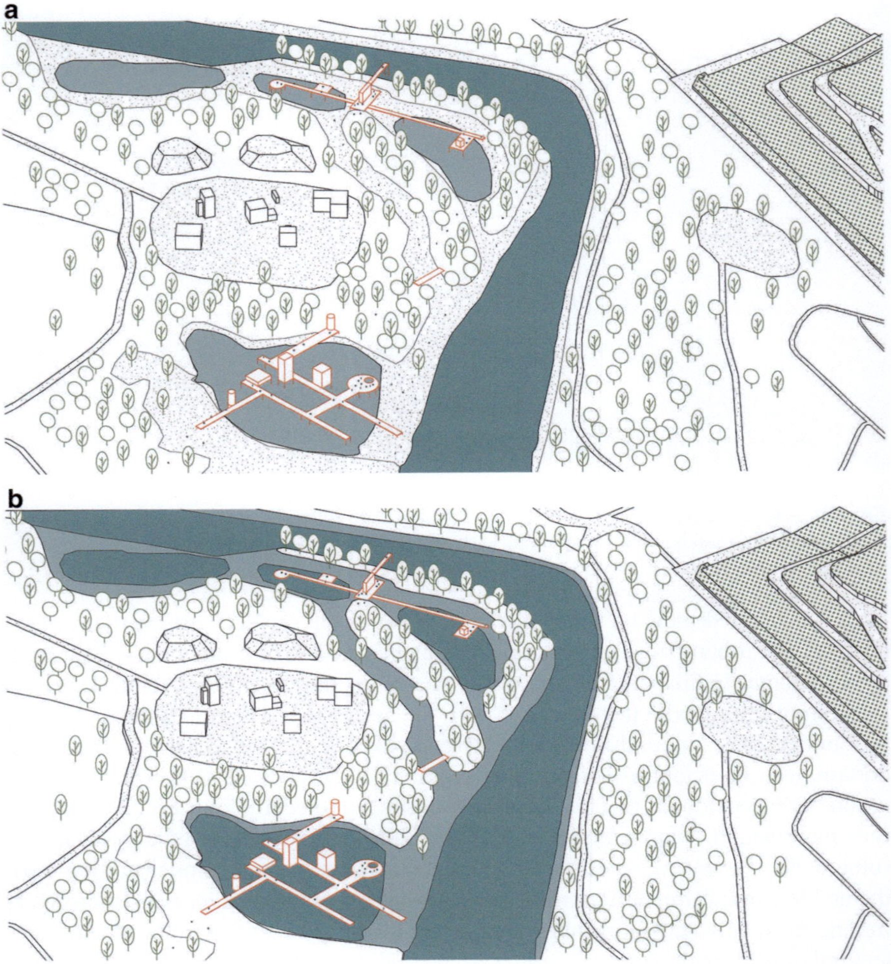

Fig. 3.10 (**a**, **b**) Detention basins and quarries, places of nature and places to stay in a different way, temporarily. Places that change their geography with the different water levels in each season or rain period (Source: author's elaboration)

for a design which makes it possible to perceive the elements into which it is divided: nature, building and infrastructure. An articulate section defines a space, which is now undefined and wasted: viability, residential/service constructions, private green areas, private or licensed allotments, agriculture, cycle paths-footpaths, area of re-naturalisation along the watercourse, in a composite but recognisable morphology.

A *Ville legére*, as it is defined by LIN in the project for *Le Grand Paris* (2013), porous (*Ville poreuse* Secchi and Viganò 2011) is an interstitial tissue made up of abandoned and unused ground and buildings in a temporary state, waiting for a vocation. The relationship with an absence of nature offers an opportunity to design spaces to make the boundaries permeable (Fig. 3.12).

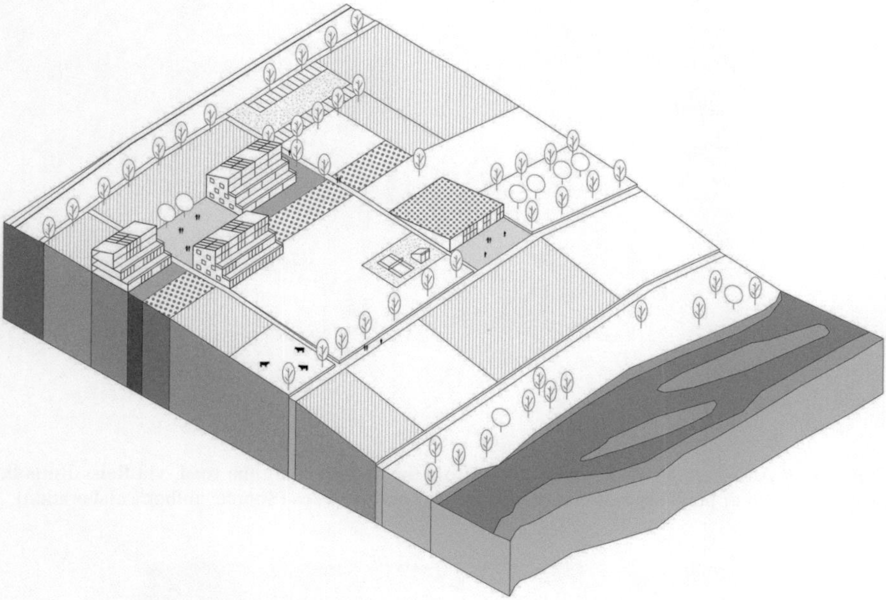

Fig. 3.11 The section represents the relations between different elements on the urban fringe on the river banks, infrastructure, buildings (housing or services), private gardens, private or in loan for use vegetable gardens, agriculture, pedestrian and bike paths, agriculture, dirt road for agricultural vehicles, re-naturalised riverbank and the river (Source: author's elaboration)

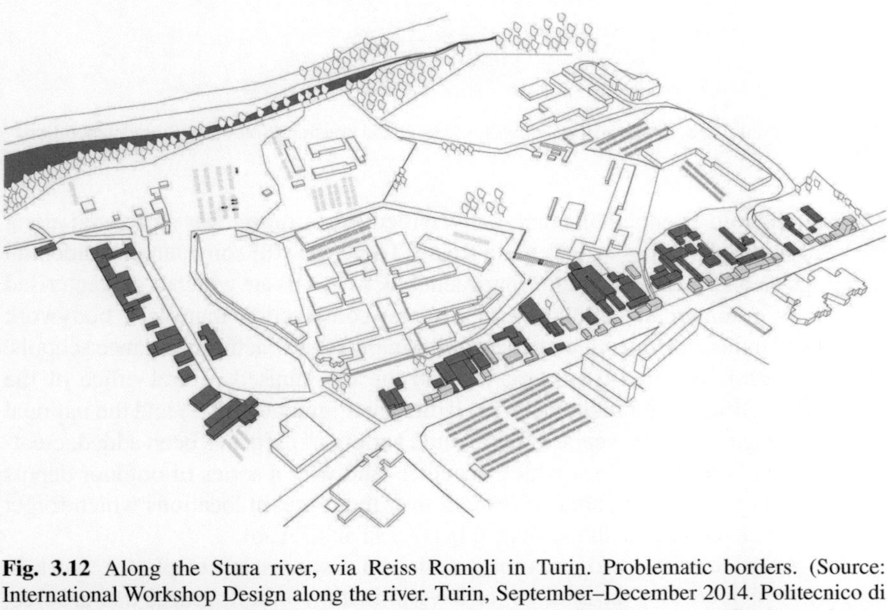

Fig. 3.12 Along the Stura river, via Reiss Romoli in Turin. Problematic borders. (Source: International Workshop Design along the river. Turin, September–December 2014. Politecnico di Torino. Tutors: Ingaramo R., Voghera A., Claps P., Mutani G., Mazzotta A. Students: Acosta Sarmiento C., Acosta Saermiento M., Carta D., Pepe G)

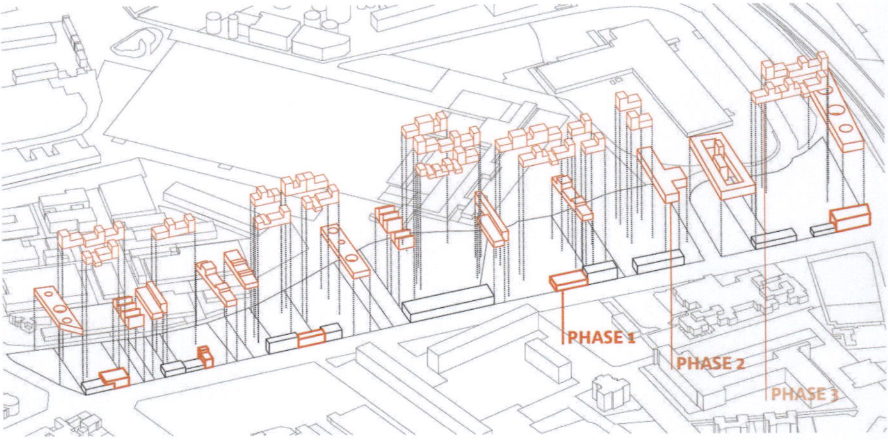

Fig. 3.13a An incremental strategy starting from the margin along the road, via Reiss Romoli, than the design of public spaces and productive and creative hubs (Source: author's elaboration)

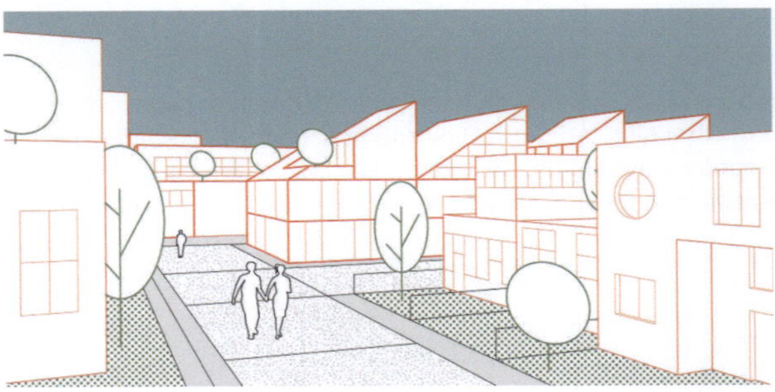

Fig. 3.13b Public and private spaces between working and housing (Source: author's elaboration)

From the mid-twentieth century, a diversified urbanisation was structured along the road that links Turin with Venaria Reale. There are still some small residential buildings on one or two floors and apartment blocks in a tissue which is characterised mainly by commercial buildings (car dealers, construction materials, bodywork shops) and industrial units interspersed with small parasite activities (dance schools, pizzerias, etc.). Towards the Stura, we find the mechanised central office of the Italian Post Office, paint manufacturers (Hutsman Pigments S.p.a.) and the national shooting range. Over the years, a fragmented variety of units has been added, creating a dilated horizontal space which occupies land with a series of outdoor depots that expands with parking areas spread all over the place, in locations which forget that they are in contact with the river (Figs. 3.13a and 3.13b).

The *design scenario* proposes intervention in a succession of phases, work in progress which enables the addition or the elimination of built spaces that are free, depending on the evolution of the demand for settlement. It isn't a finished and definite overall design but a possible vision. *Incremental design*, marked by different timing phases, combines the reuse of spaces and buildings that have been abandoned or are unused, with functional revitalisation. The transformation of the continuous margin along the road takes place thanks to the densification of the buildings, allowing the characterisation of the front, and leads to the opening of "corridors" to the river, creating a new permeability towards it. The different phases allow a dynamic use of the space, with the construction of relational spaces among the different elements: residential buildings, production units and service facilities. New residences strengthen the edge along the road, occupying unused lots, building other floors, extensions, and insertions. The weave of the design is made up of the public, semipublic and private space between the buildings and connects the spaces with a re-naturalised river. The vocation of manufacturing and services is strengthened with the creation of a space for mediation and sharing, between accommodation, manufacturing and new workshops for a creative industry.

The banks of the Stura river are not just a space for river parks either but for a pervasive structuring design where necessary, like in Bordeaux at Parc aux Angeliques by Michel Desvigne[24], the reuse of a territory which stretches for 400 hectares (Desvigne 2012) (Fig. 3.14).

3.2.5 New Living Spaces

Rapidly changing lifestyles have new needs in terms of space, dimensions and comfort, new needs in relation to the emergency of finding somewhere, even temporary, for migrants to live. This is proving to be so difficult that abusive occupation has become very common, as in the case of the Olympic Village in Turin (The Guardian, March 2016[25]). The communities are becoming more mobile and the proximity between workspace and living is often casual and temporary (Figs. 3.15a and 3.15b).

In these places characterised by fragile urbanity suspended between a productive characterisation and services, we have imagined new living spaces, which can be used for various purposes, unconventional and categorised, capable of transforming themselves. Temporariness is a characteristic that pervades the places where we work, relax and live, so architectural design has to deal with complex, evolving needs, and the theme of reuse of what already exists becomes endemic (see the PRIN national research carried out by numerous Italian universities, Re-Cycle Italy[26] and Fabian et al. 2012) (Fig. 3.16).

Space changes, altering its spatial configurations, with alternating and overlapping uses. As in the design of a narrow strip of land squeezed between the railway and infrastructure, one of the accesses to the city from the Turin-Milan motorway, which arrives from the north, crossing the river. This "in-between" space, now

[24] In Lotus International n.150.

[25] Provost and Lai (2016).

[26] http://www.recycleitaly.it/

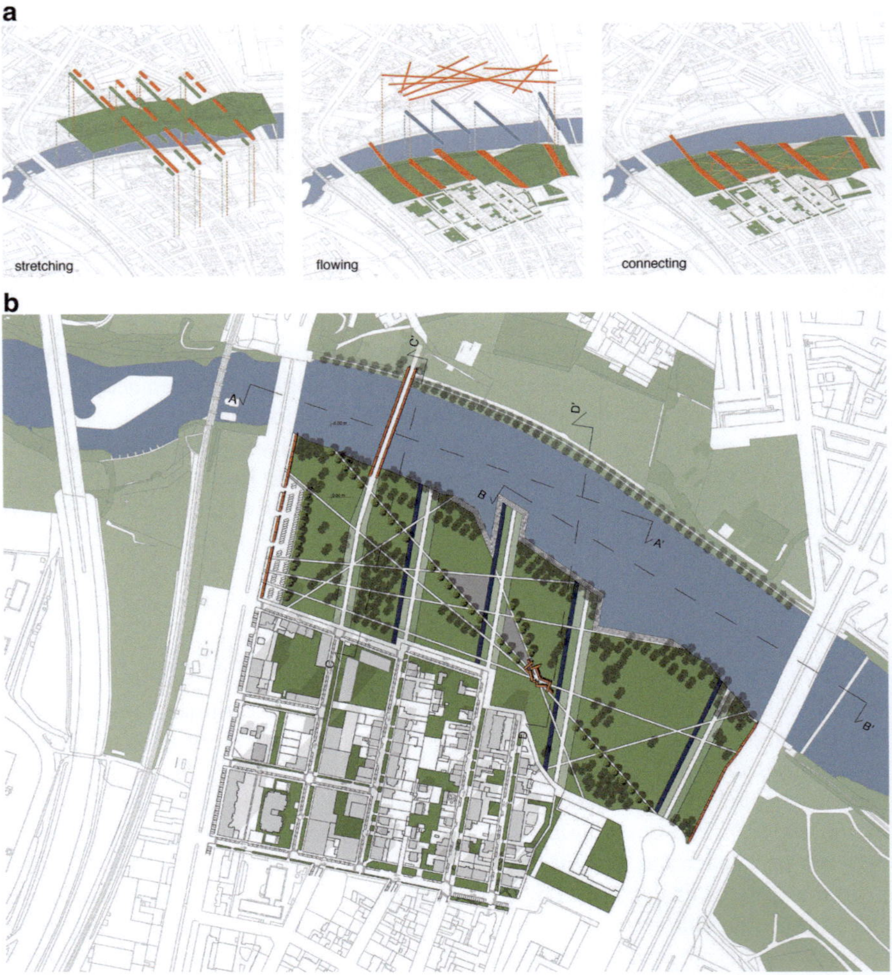

Fig. 3.14 (**a**, **b**) Permeability strategy along the Stura river in Turin. A green comb to tie again. (Source: International Workshop Design along the river. Turin, September–December 2014. Politecnico di Torino. Tutors: Ingaramo R., Voghera A., Claps P., Mutani G, Mazzotta A. Students: Catalano A., Bedello F., Scialdone A., Tassera D)

invaded by unauthorized allotments, service installations and fragmented manufacturing, becomes a place for experimenting, innovating and sharing, with continuative, diversified use, with an architecture that reuses what already exists and incorporates elements capable of structuring a permeable and attractive space, open to the rest of the district. The intersections between existing and new architectural elements define a renewed spatial arrangement, where the yard plays a role in the construction of public and private space, like in the project by the Ambrosi Etchegaray studio of Guanajuato in Mexico's Roma Neighbourhood in 2015.

Fig. 3.15a Permeable strategy to integrate industrial and commercial buildings with new housing and services. Discover the nature. (Source: International Workshop Design along the river. Turin, September–December 2014. Politecnico di Torino. Tutors: Ingaramo R., Voghera A., Claps P., Mutani G., Mazzotta A. Students: Allevato J., Battafarano I., Sandolo G)

The industrial building is no longer merely a place of work but a new reality, a creative hub with numerous functions, comprising laboratories, research centres and relational spaces, to establish relationships between the different working environments (Fig. 3.17).

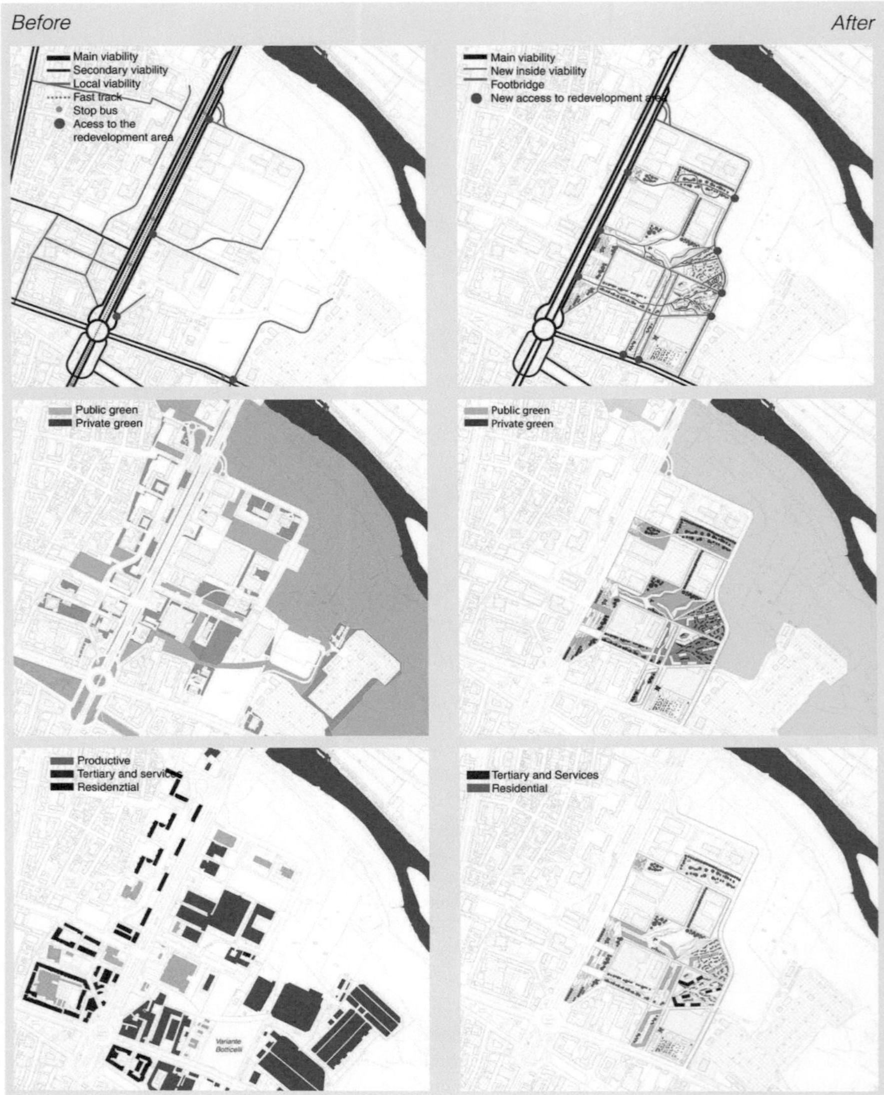

Fig. 3.15b Discover the nature. (Source: International Workshop Design along the river. Turin, September–December 2014. Politecnico di Torino. Tutors: Ingaramo R., Voghera A., Claps P., Mutani G., Mazzotta A. Students: Allevato J., Battafarano I., Sandolo G)

Fig. 3.16 Belt along Corso Vercelli on the Stura river banks (Turin). A strategy to diversify by using existing buildings integrating them with new elements able to adapt to constant change. Small specialized productions with places of sale and workplaces open to the public. Embrace the Stura experience (Source: International Workshop Design along the river. Turin September–December 2014. Politecnico di Torino. Tutors: Ingaramo R., Voghera A., Claps P., Mutani G.,Mazzotta A. Students: Alfarano E., Borra F., Pignataro A., Tempesta M)

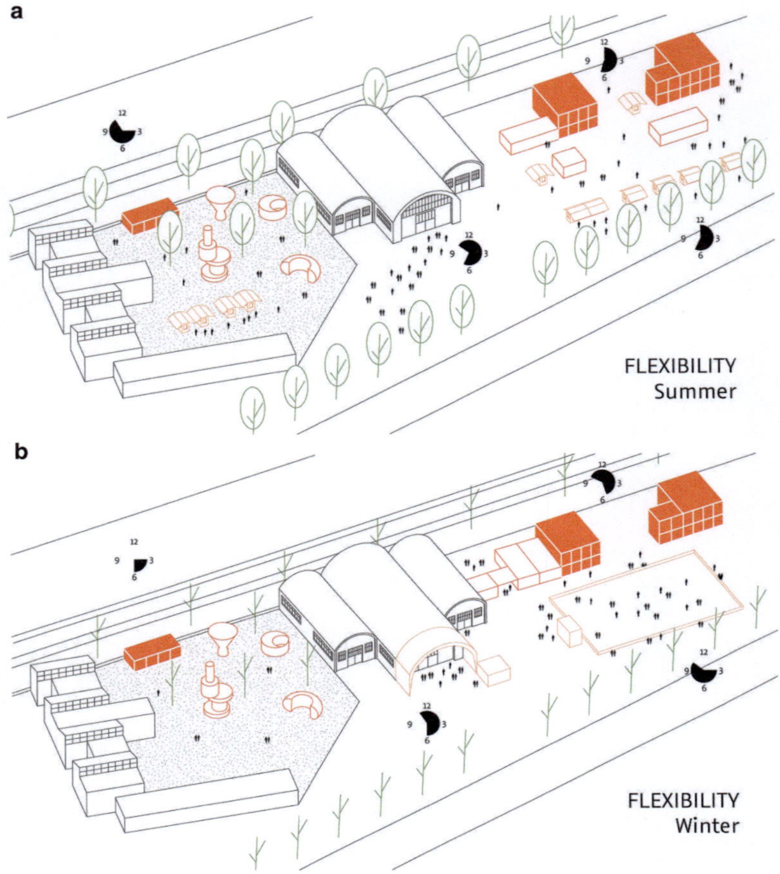

Fig. 3.17 (**a**, **b**) Different ways of living and work, flexible in different seasons and in different hours of the day. Existing buildings and new ones coexist in a dynamic and evolutive system (Source: author's elaboration)

3.3 The Effects of the Projects

These design experiences are part of a broader research experience that we have carried out in recent years in our involvement in the processes of construction of River Contracts in Piedmont (Sangone River Contract 2008–2009, Avigliana Lake Contract 2010–2011 and, in recent years, the Stura River Contract 2014–2015), promoted by what was the Province of Turin and is now the Metropolitan City and Piedmont Region. They have shown that the identification of expected landscape and local projects, proposed with alternative scenarios, has enabled the definition of shared strategies at the different scales, in support of the institutional decision and for relating to the residents and stakeholders, having political involvement and an effective role.

The component of participation in the process, which involved on-going debate with the local populations, institution and investors, with workshops and public assemblies for about 2 years, although limited to some of the users or stakeholders involved, takes on a role of political legitimacy of public decisions when administrations and governments, at a time when resources are scarce, have difficulties in defending them, especially if the choices made are imposed from above.

The Stura and Sangone river basins were used for the experimentation of large-scale policies for a long time, accompanied by urban and architectural design strategies (planning workshop from 2004 to 2010 for the Sangone and from 2005–2016 for the Stura di Lanzo): They have had different effects compared to other experimentations of the River Contracts method in Italy (Voghera A., *Territorial Design. A Characterization from Experiences*), many of which revealed a certain weakness in terms of implementation. Feasible design, set into a system strategy, as in the case of the Sangone River Contract Masterplan (Ingaramo and Voghera 2012), where the outcomes of design by parts, transformations in progress, imagined and made possible using different levels of design, are recomposed, on the other hand, has had real effects on a Grand Scale (De Rossi 2009) and also at local level in the Planning Schemes of the individual municipalities (11 municipalities were involved around the Sangone). Small municipalities, like Trana (near Turin), thanks to the project strategies theorised by the Masterplan, have been able to take part in the activities of the Corona Verde projects for the funding of local projects for river parks and for the reuse of a publicly owned property.

The research shows the validity of an approach, which allows the completion, with an unconventional process, of projects that bring together environmental and landscape issued with urban and architectural design. A vast-scale frame (see Chap. 2 Voghera) can be defined as an open weave for the placement of local projects, not as a consequence of a super-ordained strategy but as a tight structural and structuring fabric. Urban and architectural design is not a single expansion of scale but an indication to build a reference strategy and methods of comparison among the various degrees of territorial governance, stakeholders and local populations.

3.4 Roles for Design

3.4.1 Relational Design

"Now more than ever before, architecture is an architecture of relations rather than of pure objects, of dynamic relational spaces instead of static scenes…" as stated in the programmatic manifesto of the Gangemi series, directed by Renato Bocchi, *Spazio Paesaggio Architettura*.

The role of relational design, in which geomorphology intensifies the level of interdependence between nature and construction, in the search for a balance that generates spaces and architectures capable of enhancing or redesigning intrinsic potentials, emerges more clearly along the riversides. Design takes on a role of mediation between different morphological contexts, underscoring the differences and rediscovering inspiration for its determination in the intersections (Bocchi 2009).

Spaces of mediation, between those on a large sale and those at architectural level, where *intermediate places, spaces in-between* considered as privileged locations for the project of Aldo Van Eyck (1963) in opposition to the concept of spatial continuity and to the tendency to eliminate every articulation between spaces (between outside and inside, between one space and another) (Strauven 1997) , or the *threshold* by the Smithson (1968), with their streets in the air, become places for experimentation once again (Ministry of Housing Spain 2010). These are the sharing places where trans-scalar design scenarios can be implemented at landscape and local scale. If the project is the form from which to build strategies, it is in the individual fragments of urbanity, in the intermediate spaces, that the construction of a design for Gregotti's transformation into being *capable of seeing small: between things* (Gregotti 1985) takes place, between built and unbuilt, where transformation processes are experimented.

The *spaces in-between* -between the ecological networks, the different landscapes, the compact urbanisation and the river; between the buildings and the agricultural areas and the buildings themselves; but also in the same building between public and private spaces (entrances, common areas, patios and yards, terraces and greenhouses) - become important themes for a mediation project far-removed from hierarchies and classifications (Aymonino 2012). They propose new spaces for living, working and relaxing, and sometimes these are interchangeable, offering relational opportunities and relating to one another in an eco-sustainable dimension.

These places are a fixed, "suspended" capital of our territories (Mazza 2004), made up of natural and tangible components, open contaminated spaces that are unused or underused, abandoned buildings, places transformed by man's work, awaiting a vocation. Even ordinary buildings take on the role of assets, in that they are part of a "consumed" ground. In Italy, more than elsewhere, since the introduction of the new law on ground consumption[27], the capitalisation process has dilated its sense and become an opportunity for design based on adaptive reuse which envisages an adaptation of new content to the existing container. Rejected, underused and contaminated elements are possibilities for design which regenerates, transforming living spaces to suit different scales, with morphologies that second the natural geographies: differences in height, areas where flooding occurs, biodiversity.

The scale of the landscape project interweaves with architectural design, in a logic which focuses on re-naturalisation and reclamation, in the reconstruction of ecological networks as the framework of a design of territory detached from landscape-related nostalgia. This design, which sees *a nature that is almost reified in its conservation of the environment* (Gherardini and Olmo 2015), stems from the interpretative rewriting of a landscape and of the artifices that are part of it,

[27] On 22 May 2016, the Chamber licensed the proposed Law to restrict ground consumption, with the aim of zeroing national ground consumption by 2050. In the past fifty years, a million and a half hectares of land have been made impermeable.

incorporating local projects, sometimes of minute entity, which touch materials requiring adaptation, redefinition and relation to one another, via urban and ground design (Secchi 1986, 2011) which leaves possibilities and choices for present and future uses.

3.4.2 Design Strategy Through Scenarios

Aldo Rossi, who supported a disciplinary autonomy in the 1960s, attributing the verifiability and the tangibility necessary to architecture for it to represent a more complex reality, later wrote (in 1976), in Lotus International, that:

> (…) technical man and/or artist must offer alternatives to the growth of cities so that these alternatives can be discussed, understood and hence either accepted or rejected by the people who live in those cities. Gone are the days of urban models and with them also has gone the age of urban techniques, self-description, and functions passed off as solutions. The city must be dealt with each time, by gathering and developing its contradictions, day by day, directly." He underlines the failure of the Italian Piani Regolatori (Planning Schemes) never put into practice, destined to perform a sort of penitential function through a zoning which has never been seriously carried out. (Rossi 1976)

Now, like then, but in a clearer way at a time of crisis which requires a new way of acting, different design alternatives emerge as needs in the definition of our cities, *scenarios* which offer spatial solutions that we can compare and discuss, which are incremental and open to alteration. The *scenario* becomes spatial implementation of evolving interscalar strategies capable of imagining and directing specific choices that concern, first and foremost, people's quality of life, as the Horizon 2020 European framework programme wish to suggest.

An open city can *be well designed*, as claimed by Sennet, using three strategies that are generated on the principle of a lack of determination. *These designs involve creating ambiguous edges between parts of the city, contriving incomplete forms in buildings and planning for unresolved narratives of development* (Sennet 2006).

3.4.3 Incremental Design

An architectural and urban design proposes alternatives and is implemented at different times, with time being a major factor (Carmona 2014; Schurch 1999). However, in the age of Twitter, there are possible intervention strategies that require long implementation times, which second contingent financial resources or the changed conditions of transformation, becoming a tool for knowledge and confrontation between public administrations, investors and the population. In the experiences of a designer, projects are often implemented in lots, diluted over a period of time, in order to source funds or await the arrival of funding in instalments, but this is not our case. We are talking about a project which is built up and changed over time, which proposes a scenario to aim for but which does not necessarily have to be observed in absolute terms, because it defines an optative rather than an imperative form, opening up the way to spatial modifications of the context and of the container and its content. These are not necessarily solutions to the problem as a whole, but to minute problems, with local forms and dimensions that can also be detailed but capable of having a considerable influence on system strategies and designs. The transformation of even just one building in an urban boundary area, or of an infrastructure which provides access to the other bank of a river, not merely as a crossing but as a conquered space, where we can spend time and discover new views (Mehta 2014), a new centre within the district (see the house in the park at Mirafiori, Turin), is a generator of incremental processes of regeneration. This does not mean that the designer doesn't have to deal with large-scale projects but that there are a wide range of possibilities that are often more feasible, with a greater impact on the quality of life of those who live in these places.

3.4.4 Participatory Design

If it is true, as Bauman (Bauman and Bordoni 2015) argues, that the stagnation of the economic crisis we have been experiencing for several years is the result of a divorce between power and politics, it is clear that the state's inability to manage the change, in Italy more than anywhere else, has resulted in a stabilisation of the crisis that some scientists, like Bordoni, believe can continue forever. Under these conditions, design takes on the role of concerting tool, capable of restoring the capacity for prefiguration to those who govern our territories, in agreement with those who live in them, for a radical rethink of the processes that lead to the choices of its forms.

Not a definitive, prescriptive project but open opportunities for prefiguration, the fruit of concerted actions that cover the various scales of the project, which, as such, offers morphologies rather than standards, possible updates and structural variants, in a concerted system in which co-planning is a need and not just an obligation, as is communication with residents (Magnaghi 1998). Small, localised and strategic operations, not too far removed from urban acupuncture (De Solà-Morales 2008), can be capable of triggering even partial and minute regeneration processes, made feasible also by processes of involvement of a specific "part" (Martinotti 1993), of a group, whatever its size, of citizens or users which has little to do with the rhetoric of generalist participation. This is a shared transformation, distinguished from the practices of participation in the 1970s (De Carlo 1972, 2002) expanding its sense. We "design by attempts" to reveal the situations and open up new ways to their transformation, De Carlo wrote, reiterating his convictions once again in 2002, and for those who raised doubts as to the end of architecture, he indicated the need for an architecture that has to directly or indirectly involve those who use it in order to survive, becoming a bearer of instances, needs and interpretations that only an architect can and must organise and render operational.

More and more often, projects imposed from above are contested and blocked by committees of residents or local groups, as in the case of the Mediaspree in Berlin, where a group of citizens, artists and experts in various sectors concerted their efforts to fight the demolition of disused industrial buildings and the construction of buildings and offices. They work with architects, who take on the role of interpretative mediators, on practices like that of the Holzmarkt[28], a district "for everyone", with an ecological fair imprint, spatially defined by a low-height and low-density architecture. Groups like those in Spain (Esto no es un solar in Saragoza), cooperations (like ALinha in Lisbon), platforms (Vivero de Iniciativas Ciudadanas) or groups supported by research departments (like Temporiuso in Milan) joined traditional design processes, with brilliant results, as in the case of the now iconic High Line of New York.

3.4.5 Design for Living Spaces

If the environmental question is one of the main engines that drive design today, it will become even more crucial in the future (Barbieri 2016). For architectural design it is an opportunity to regain its own ethical dimension as a social service which does not envisage marked destinies, in which objectual *authoriality* falls back, prioritising the *choral strategy*, in which construction becomes part of a more articulate scenario, part of a whole in which it finds its own dimension of change.

The themes of landscape and territory took a leading place in research performed in various fields from the 1990s and primarily among architects involved in design activities; in Italy, way back in the 1960s, in the monographic issue of Edilizia Moderna, Gregotti wrote about *Forma del territorio*. The idea of a systemic project (Aymonino 2012) places the role of the design of the city and architecture under discussion, shifting attention from the objectual form to the way in which architecture is related to the space; the relational systems acquired by the built elements are fundamental to the projects, involving the skills of different disciplines. Multidisciplinary teams become commonplace in an approach to design that leans increasingly towards a substantially trans-scalar character, especially when the project takes the form of operational research, as in the Atelier du Grand Paris (Atelier International du Grand Paris 2013). The architect is not a specialist. An architect creates relations, designing within a compromise which is far removed from conformism (Siza 2008) and a project completely detached from *prescriptive rhetoric* but which finds a sort of feasibility in the dimension of *communication and concertation*, becoming a tool for building change (Gregotti 1984).

If, as emphasised by Aravena in the press conference for the opening of the 2016 Venice Biennial, Reporting from the front, the sense of design lies in improving the quality of life, the rediscovery of an ethical role of the dimension of the project takes

[28] http://www.holzmarkt.com

us back to the social mission of architectural design. Not the design of exemplary models, but strategies for everyday living: interpreting, accompanying, proposing and imaging spaces for life.

References

Atelier International du Grand Paris (2013) Habiter Le Grand Paris. Archibooks et Sautereau Éditeur, Paris

Aymonino A (2012) Più spazio meno volume: un racconto in movimento. In: Ricci M (ed) Nuovi Paradigmi. LISt Lab, Trento

Barbieri P (2006) Infraspazi. Meltemi, Rome

Barbieri P (2016) Geo Città? List, Laboratorio Internazionale Editoriale, Trento

Bauman Z, Bordoni C (2015) Stato di Crisi. Giulio Einaudi Editore, Turin

Bertoncello B, Dubois J (2009) Marseille Euroméditerranée. Accélérateur de Métropole. Parenthèses, Marseille

Bianchetti C (2007) Urbanistica e sfera pubblica. Donzelli, Rome

Bianchetti C (2012) https://territoridellacondivisione.wordpress.com/2012/10/29/la-confluenza-turin-a-desputed-territory/#more-2041, August 2016

Bocchi R (2009) Progettare Spazi e il movimento. Gangemi Editore, Rome

Burgos F, Ginés G, Porras-Isla F (eds) (2014) Landscapes in the city. Madrid Rio: geography, infrastructure and public space. Typsa, Madrid

Caccia Gherardini S, Olmo C (2015) Architecture and heritage. In: Bianchetti C, Cogato Lanza E, Enver Kercuku A, Sampieri A, Voghera A (eds) Territories in crisis, architecture and urbanism facing changes in Europe. Jovis, Berlin

Carmona M (2014) The place-shaping continuum: a theory of urban design process. J. Urban Des. 19(1):2–36

Carter DK (ed) (2015) Remaking post-industrial cities. Lessons from North American and Europe. Routledge, London

Chasseriau A (2008) Au cœur du renouvellement urbain nantais: la Loire en projet. Norois 192 (2004/3)

Chemetoff A, Berthomieu JL (1999) L'Île de Nantes, le plan guide en projet. Éditions MeMo, Bureau des Paysages, Nantes

Clement G (2005) Manifesto del terzo paesaggio. Quodlibet, Macerata

Clementi A, Cao U, Carnevale G (eds) (2014) Scoprire il Fiume. Recovering the Jiu Qu River. List Editore, Trento

Corboz A (1985) Il territorio come palinsesto. Casabella 516:22–27

Cuff C, Dahl PJ (2015) Housing in the RiverCity: rethinking place and process. Mistra Urban Futures, Göteborg

De Carlo G (1972) An architecture of participation. Royal Australian Institute of Architects. Melbourne Architectural Papers, Melbourne

De Carlo G (2002) Sulla progettazione partecipata. In: Scavi M, Romano I, Guercio S, Pillon A, Robiglio M, Toussaint I (eds) Avventure Urbane, progettare la città con gli abitanti. Elèuthera, Milan

De Rossi A (2009) Grande scala: architettura, politica, forma. LISt Laboratorio Internazionale Editoriale, Trento

De Rossi A, Durbiano G (2006) Torino 1980/2011. La trasformazione e le sue immagini. Allemandi, Torino

Dematteis G, Governa F (eds) (2005) Territorialità, sviluppo locale, sostenibilità: il modello Slot. Franco Angeli, Milan

De Solà – Morales M (2008) A matter of things. Nai Publishers, Rotterdam

Desvigne M (2012) The landscape as precondition. Lotus Int. 150:20–27
Dossier Marseille (4 février 1994) Le Moniteur, pp 93–105
Durbiano G (2014) Etiche dell'intenzione. Christian Marinotti Edizioni, Milan
Epaem (2006) Euroméditerranée 10 ans. Plaquette de communication, Marseille
Fabian L, Giannotti E, Viganò P (eds) (2012) Recycling city, life cycles, embodied energy, inclusion. Giavedoni editore, Pordenone
Grahame Shane D (2011) Urban design since 1945. A global perspective. Wiley, Hoboken
Gregotti V (1984) Modificazione. Casabella 498-499:2–7
Gregotti V (1985) Dieci Buoni consigli. Casabella 516:2–3
Ingaramo R, Voghera A (2012) Planning and architecture. Searching for an approach. Alinea Intenational, Florence
Mack L (2014) Ideas flow through the river, too. Star Tribune, February 6
Magnaghi A (ed) (1998) Il territorio degli abitanti. Società locali e autosostenibilità. Dunod, Milan
Martinotti G (1993) Metropoli, la nuova morfologia sociale della città. Il Mulino, Bologna
Mazza L (2004) Piano, progetti, strategie. Franco Angeli, Milan
Mehta V (2014) The street. A quintessential social public space. Routledge, New York
Ministry of Housing Spain (2010) Social housing & city
Molga P, Couartou M, Sbriglio J (2014) Marseille Euroméditerranée. Un Laboratoire de l'architecture du soleil. Autrement, Paris
Ota K (2004) Watertrees. Toyo Ito: Parque de La Gavia, Madrid. Domus 868:30–39
Palazzo D (2008) Urban design. Mondadori Università, Milan
Plowright PD (2014) Revealing architectural design. Methods, frameworks and tools. Routledge, London
Provost C, Lai S (2016) Occupy Turin: refugees find a home in Italy's abandoned Olympic Village. The Guardian (City), 2 March 2016
Rossi A (1976) La città analoga. Lotus Int 13:7–8
Samson M (1999) A Marseille, Euroméditerranée invente la ville ancienne sur la ville neuve. Le Monde, 15 Octobre
Schurch TW (1999) Reconsidering urban design: thoughts about its definition and status as a field or profession. J Urban Des 4(1):5–28
Secchi B (1986) Progetto di Suolo. Casabella 520-521:19–23
Secchi B (2011) Come ripensare il progetto urbano. Casabella 797:100–102
Secchi B, Viganò P (2011) La ville Poreuse. Un projet pour le Grand Paris et la métropole de après Tokyo. Metispresses, Paris
Sennet R (2006) The open city, urban age. Newspaper Essay, November, Berlin
Siza Á (2008) Sulla pedagogia. Casabella 770:3–5
Smithson A (ed) (1968) Team 10. Primer Studio Vista & Reinhold, New York
Strauven F (1997) Aldo van Eyck, the shape of relativity. Architecture & Natura, Amsterdam
Van Eyck A (1963) Beyond visibility, about place and occasion, the in between realm, labyrinthian clarity. Situat Times 4:79–85
Venezia F (1984) La costruzione del limite: un piano per Lauro. Casabella 498(9):100–105

Chapter 4
Experimenting Design Tools

Roberta Ingaramo and Angioletta Voghera

Abstract The work we have carried out in relation to the Action Plan of the River Contracts has been an opportunity for communication between urban planning and architectural design, to build up a territorial project. Major topics identified and a methodological proposal, the *integrated multi-scale projec*t, are discussed. The method uses a tool to provide decision-making support and an operational technique: the *Action Agreement* and the *transecting section technique*. Different participated design experiences are described to illustrate the evolution of the research.

Keywords Integrated multi-scale methods • Design topics • Transecting sections • Action agreement • Architectural design • Urban design • Landscape design • Participation

4.1 On Design

The territorial project is one of the aims of the Action Plan of the River Contracts (see Chap. 2 Voghera and Chap. 6 Berruti and Moccia), which defines vast area transformations of urban, farming, natural and rural places. All too often, the project is the "*chaotically stratified result* of the sum of sectorial projects that functionally occupy regional space, activated by specific interests. More often than not, it is contradictory and conflictual. At best, it is vaguely "regulated" by plans and public policies" (Magnaghi 2014, p IX). Magnaghi also says (2014, p X) that the territorial project is "aerial matter, poorly anchored to experimental verification, still lacking a multidisciplinary scientific structure, when compared to the disciplinary expertise

R. Ingaramo (✉)
Department of Architecture and Design, Politecnico di Torino,
Viale Mattioli 39, 10125 Turin, Italy
e-mail: roberta.ingaramo@polito.it

A. Voghera
Interuniversity Department of Regional and Urban Studies and Planning,
Politecnico di Torino, Viale Mattioli 39, 10125 Turin, Italy
e-mail: angioletta.voghera@polito.it

R. Ingaramo, A. Voghera (eds.), *Topics and Methods for Urban and Landscape Design*, Urban and Landscape Perspectives 19,
DOI 10.1007/978-3-319-51535-9_4

of architectural design or 'urban design', historically the domain of architecture and city planning (at its origins as a discipline)".

Urban design, often referred to in Italy as urban planning or city project, has become increasingly important over the past 30 years. The term urban design, which was first used in the fifties, after the Second World War, initially by a group of professionals operating in the sector, is now used to indicate almost everything that concerns human settlements.

The term was used by Patrick Abercrombie and John Forshaw, in their *County of London Plan* to rebuild the city after the war. Then, in 1953, Frederick Gibberd approached the manual prototype with *Town Design*. Abercrombie realised the need for a disciplinary field that covered both the disciplines of design and planning, capable of constructing territorial design (Grahame Shane 2011).

Besides, the extension of the concept of spatial pertinence of architectural research brings an increasingly substantial relationship with similar disciplines into play. These disciplines focus on environmental and ecological problems that have a considerable influence on the morphological definition of the spaces that we live in.

And this is mainly what caused us to start thinking about how to effectively ensure communication between urban planning and architectural design to build up a territorial project (De Carlo 2008), in support of the river agreements, effective on a large scale as well as at architectural level (Ingaramo and Voghera 2012). In fact, urban planning has always focused on defining large-scale scenarios so that strategies can be adopted as part of territory and landscape planning and programming tools (Mazza 2003). It tends to address these issues by concentrating in particular on policies, rules and institutional and social actors; however, it should now reflect, as it did in the past, on the implementation, direction and repercussions of these projects. Since research now includes large-scale and landscape issues (CoE 2000, 2008; Voghera 2011), architectural and urban design often tends to ignore or underestimate the complex strategies, stakeholders and actors involved in the transformation of territorial areas and habitats; instead, it prefers "image" projects that appear to guarantee quality and social and institutional consensus. With this in mind, we tried to redefine a design methodology in order to establish dialogue between several disciplines. It should explore new settlement patterns and flexible types of architecture for inclusion in the territorial system; it should introduce local policies, programmes, design capabilities and experimentation in specific areas, into the strategic framework of the Masterplan. This kind of experimentation has been developed by *transecting sections* – three-dimensional images of the landscape (see paragraph 4.3 and Chap. 3 Ingaramo) and details of architectures – located in areas and sites of specific interest. These are strategic locations for the urban service system, ecological needs and cultural and historical heritage; these strategic sites can also be turned into focus areas in order to enhance the entire system.

4.2 Within Participated Designs

The Sangone and Stura River Agreements are opportunities not only to experiment with innovative design approaches but also to define territorial and landscape transformation scenarios, affecting different areas of design at territorial level, such as wide areas, borders and focus areas.

First of all, the large area river basin and its territory require design strategies to reinforce landscape values: the river, its waters, banks, ecological/environmental potential, historical/cultural heritage, rural areas, identity-related symbols and traditions, borders and inhabited areas. These strategies also have to consider local planning capabilities and the economic resources required for implementation.

Secondly, the borders between different kinds of territories have to be redesigned and characterized, starting with the urban fringe areas which infiltrate the rural and recreational greenbelts. This kind of design has to propose something that will evolve and adapt to suit the growing contemporary demands of the territories, whether they being environmental, socio-economic or cultural.

Thirdly, the focus areas require the elaboration of protection and enhancement projects for the trans-scalar experimentation of integration between new ecological and fruition networks. These projects should preserve excellent, ordinary, natural, rural and urban areas and landscapes, while transforming inhabited or underused areas with new sustainable, exportable, settlement types and models.

To experience these trans-scalar designs (from wide to local scale), we based planning and design on the construction of processes for participation in forums, workshops and meetings, intended to help establish shared territorial development and landscape enhancement perspectives and, in so doing, influence the projects.

The workshop is an analysis and design tool, an opportunity for comparison and experimentation of defined themes, which reflects the consolidated nineteenth century tradition of the *charrette* (from 1830, the date on which the *École des Beaux-Arts* took up permanent residence in 14 Rue Bonaparte), which the students at the *École* in Paris had to work to, also revisited by New Urbanism.

In both architectural schools and in the professional sphere, the workshop has become one of the consolidated methods of tackling design, especially trans-scalar and interdisciplinary design. Used at specific times, it is an effective way of sharing choices, making them socially acceptable, using design as a powerful tool for building knowledge and decisions. Aside from the rhetoric of participation, it can be used to make institutional and social actors (public and private; see Contract Agreement; paragraph 4.3) aware of their responsibilities.

So our participation process in those cases is founded on an Interdisciplinary Design Workshop involving different groups of students:

- For the Sangone River Contract, the workshop involves students of Architecture from Polytechnic of Turin and of Natural Sciences and Agriculture. It was activated to establish a Masterplan for the Action Plan of the Sangone River Basin River Agreement (Ingaramo and Voghera 2009a, 2010), aimed at solving several interrelated problems, enhancement of overall biodiversity and the footpath

Fig. 4.1 A complex strategy for the design of the no man's land, implicating the gradual transformation of degraded and underused landscapes in time. Embrace the Stura experience (Source: International Workshop Design along the river. Turin September–December 2014. Politecnico di Torino. Tutors: Ingaramo R., Voghera A., Claps P., Mutani G.,Mazzotta A. Students: Alfarano E., Borra F., Pignataro A., Mattia Tempesta M)

network, promotion of fruition and accommodation facilities, establishment of new activities and the redesign of urban boundaries, in an attempt to identify a new settlement model not restricted to the definition of a building type. The methodology is based on the merger of urban planning and architectural and urban design, biological and natural sciences and landscape ecology, considering different environments at the same hierarchical level and integrating actions on vast and local scales.

- For the Stura River Contract, the workshop involves planning and architecture students, oriented by a team of experts in agriculture, energy, hydrogeology, urban planning, architecture and natural and built technologies. The groups focus on defining a strategic scenario and local projects, relating to the River Contract.

The Sangone and Stura designs are not so much cases of urban requalification or regeneration, as reclamation and refurbishment, strategies capable of triggering complex processes for the redefinition of the layouts of the urban landscape, also attracting the consensus of the local people and the approval of the public administrations (see the following figures related to the latest Stura Interdisciplinary Workshop in Turin) (Figs. 4.1, 4.2, 4.3, 4.4, 4.5, and 4.6).

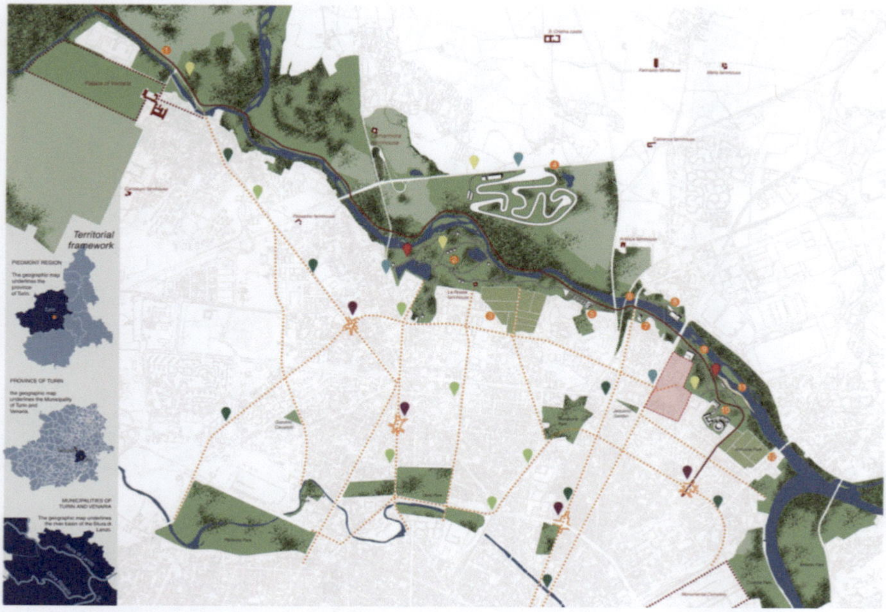

Fig. 4.2 A vision for the social reappropriation of the quarry territories and a vision relating to the river as a territorial park. Discover the nature (Source: International Workshop Design along the river. Turin, September–December 2014. Politecnico di Torino. Tutors: Ingaramo R., Voghera A., Claps P., Mutani G., Mazzotta A. Students: Allevato J., Battafarano I., Sandolo G)

4.3 Design Issues

The work we have carried out highlights the need to design the river territory with a transversal vision capable of incorporating all the relationships among the natural, rural and urbanised areas of the river at basin scale. This brings portions of land that can sometimes be some distance from the river course into the equation, reducing spatial, visual and symbolic relational systems. A project which is strongly supported by mutual relations between the territories engaged in the RC. The design of these spaces highlights the different characteristics of the materials used in their construction, starting with the definition of precise physical limits (nature/buildings, park/countryside, park/urban public space). The aim is to convey strong and immediate recognisability to spaces, which currently seem blurred, lacking defined character and recognisable identifying features. This is an activity which enables that reclamation of places required of design also by the European Landscape Convention.

The identification of strategies capable of redesigning individual locations is necessary to physically redefine territories with transversal actions at basin level, explaining and configuring sharing places. This is achieved via a relational approach which goes to work on what already exists, reconfiguring it with operations that can

Fig. 4.3 A vision of gradual conversion of industrial areas to create a linear river park, linking river and historical nodes and creating new man-made wetlands for ecological water depuration. Green line – river, networks, park (Source: International Workshop Design along the river. Turin, September-December 2014. Politecnico di Torino. Tutors: Ingaramo R., Voghera A., Claps P., Mutani G, Mazzotta A. Students: Bello C., Comoglio E., Destudio M., Nuzzo F., Sandrone I)

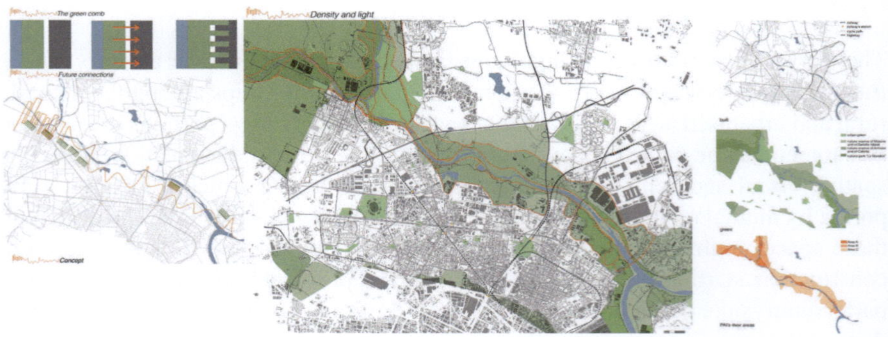

Fig. 4.4 A complex territorial strategy that will redefine the margins of natural, urban and rural areas. The strategy consists in a gradual transformation process for contaminating urban fringes, changing their image and recreating public and private spaces. The strategy takes rundown mixed production areas, imagining their refurbishment and a gradual system of operations to fill the voids, create new building surfaces and lighten up the density of the city with the installation of green areas. A green comb to tie again (Source: International Workshop Design along the river. Turin, September-December 2014. Politecnico di Torino. Tutors: Ingaramo R., Voghera A., Claps P., Mutani G, Mazzotta A. Students: Catalano A., Bedello F., Scialdone A., Tassera D)

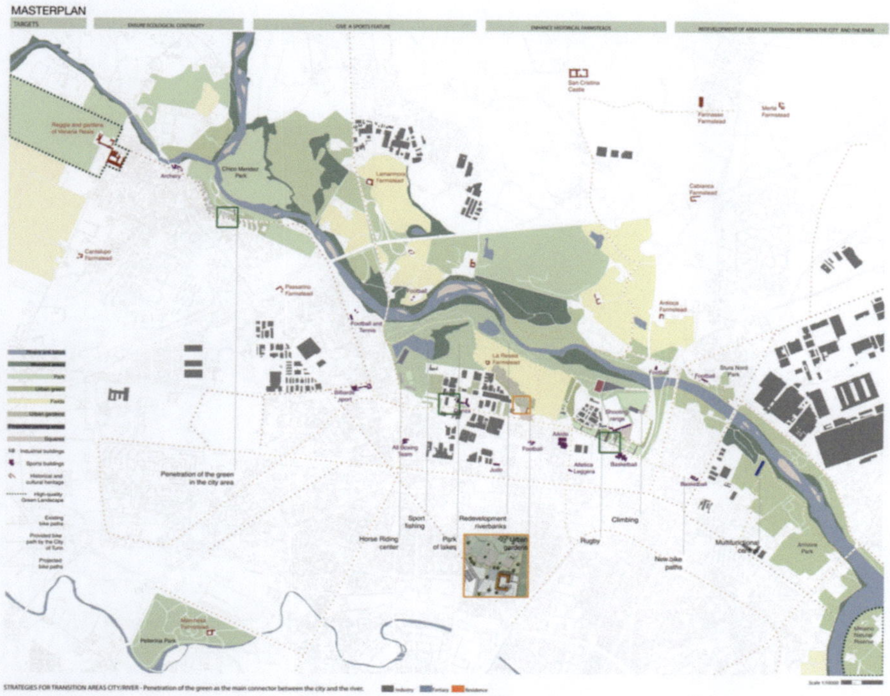

Fig. 4.5 A park for the fruition of different landscapes: the project consists in minimal interventions for the fruition of different landscape areas which change with the seasons. Problematic borders (Source: International Workshop Design along the river. Turin, September-December 2014. Politecnico di Torino. Tutors: Ingaramo R., Voghera A., Claps P., Mutani G, Mazzotta A. Students: Acosta Sarmiento C., Acosta Saermiento M., Carta D., Pepe G)

be carried out on a capillary, but also systemic, basis, of selective densification or liberation of strategic areas, with a view to ecological enhancement, for public use or fruition. An alternative could be the redefinition of the connective system required to enhance nodes, areas, territories and underused landscapes, reinforcing those few identifying elements present within these territories, identified thanks to the participated planning process implemented.

So we identify some major topics relevant to the design of these areas, which are important in the research currently being carried out on the city and are useful for student work: density and lightness, links and connections.

Each design activity, whether pertinent to local contexts or the vast area, was discussed with the stakeholders on various occasions: during the initial definition of the goals (in September 2014), during operational activities (full immersion week) with an evening *charrette* discussion, and during the once-a-week work sessions in the following months (October–December). The results were discussed at the final convention held in December, and prizes were awarded by the international panel of experts (see Chap. 8 Sabaté Bel, Chap. 12 Palazzo and Diko, Chap. 13 Plowright).

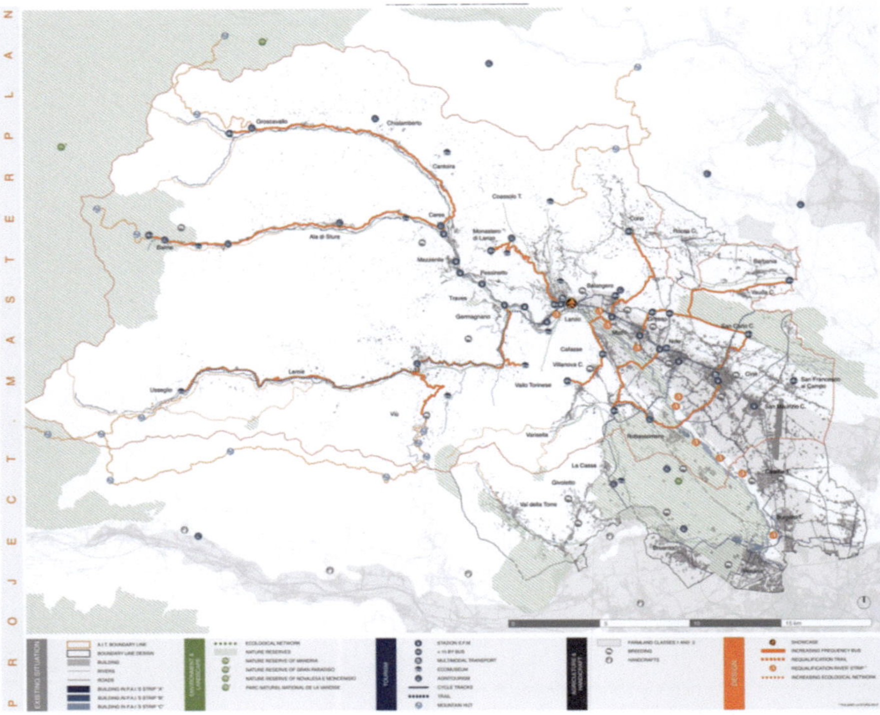

Fig. 4.6 A strategy for the enhancement of natural, economic, cultural and recreational nodes in the Lanzo Valley, to help develop tourism in these times of financial crisis. Showcase of "Valli di Lanzo" (Source: International Workshop Design along the river. Turin, September-December 2014. Politecnico di Torino. Tutors: Ingaramo R., Voghera A., Claps P., MutaniG, Mazzotta A. Students: Costamagna E., Pollini S., Torchio G., Tania Comelli T., Emilio Nervi E)

4.3.1 First: Density and Lightness

Density and lightness are normally seen as two opposing concepts, but now, more than ever before, they are subject to reflection on the successful completion of the dichotomy (like *ville légère*, as the LIN project team for Grand Paris 2013 called it; see Chap. 3 Ingaramo). Most of the areas along the Stura river seem to fall under this definition of lightness.

They are a soft part of the urban scenario, often with a single function (environmental, industrial, commercial, service, green spaces under management or abandoned land awaiting disposal). So lightness can be seen as a resource, an important concept to be worked on in the project for the city in question: increasing permeability, flexibility and diversity but also strengthening connectivity, creating and redefining spaces (Fig. 4.7).

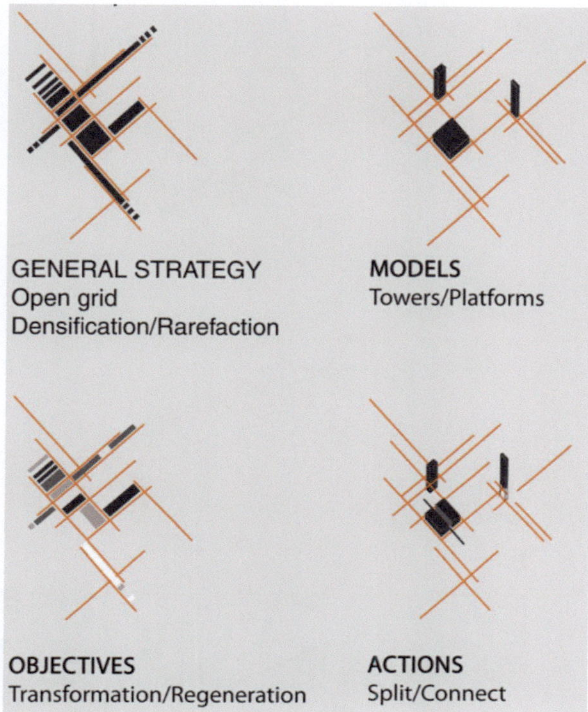

Fig. 4.7 Density and lightness along the Stura, in Venaria and Turin. Discover the nature (Source: International Workshop Design along the river. Turin, September–December 2014. Politecnico di Torino. Tutors: Ingaramo R., Voghera A., Claps P., MutaniG, Mazzotta A. Students: Allevato J., Battafarano I., Sandolo G)

4.3.2 Second: Links and Connections

The second theme looks at links and "connections". On one hand, the idea of strengthening the system can be tackled as a link, an enhancement of ecological functionality, a network of nodes, infrastructures and landscape and historical, functional and economic values. On the other is the theme of building the city as a continuum, with the reinterpretation of places with a single function (commercial, industrial, infrastructural), thanks to the physical and functional refurbishment of the spaces and the assignment of new urban values.

A popular theme in this direction is the reconstruction of margins, and it is strategically important in relation to the mirror vision of the relationship between city/river and river/city: a meeting place between different recognisable situations, a border space from which to launch the project. Is it better to compact the buildings as a complete and hard margin or to improve permeability, redefining the voids and creating new penetrations in the consolidated tissue? (Fig. 4.8).

These thoughts implicate the need to think of a different way of living (in industry, the rural context, temporary, etc.) and of production venues and of conception as creative hubs. The industrial building is no longer a simple workplace but something completely new, a creative hub, a multifunctional location with laboratories,

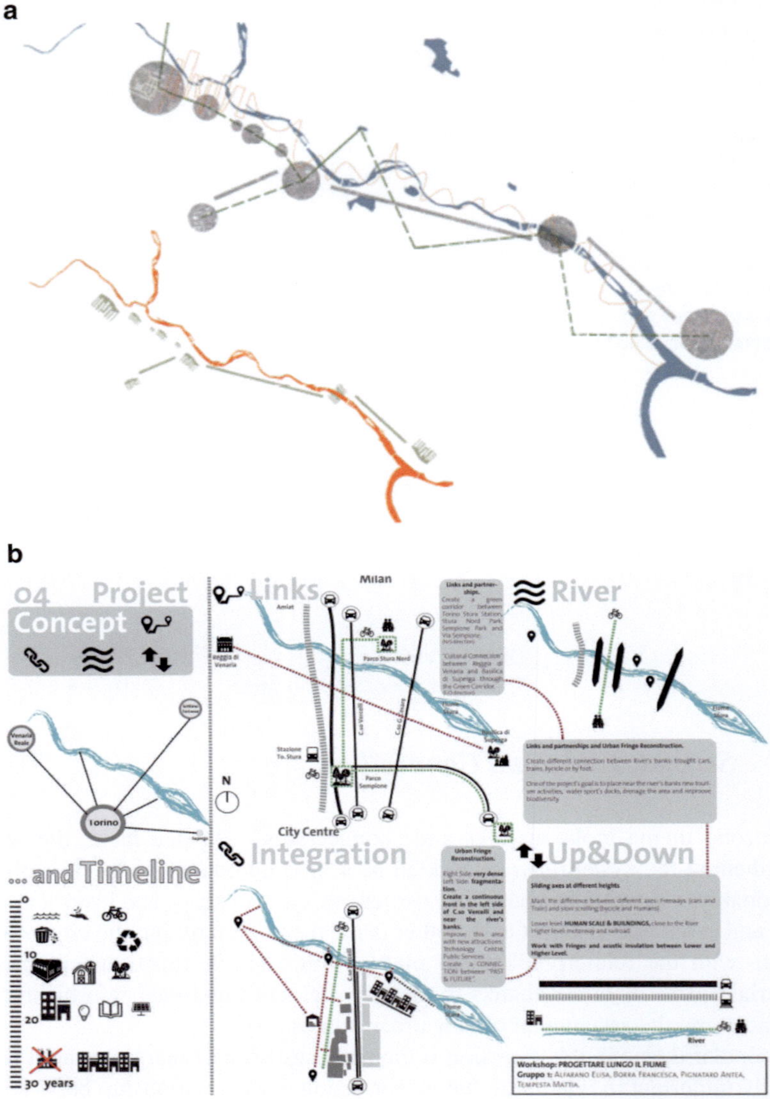

Fig. 4.8 (**a**, **b**) Links and connections. (**a**): A *green* comb to tie again (Source: International Workshop Design along the river. Turin, September-December 2014. Politecnico di Torino. Tutors: Ingaramo R., Voghera A., Claps P., MutaniG, Mazzotta A. Students: Catalano A., Bedello F., Scialdone A., Tassera D.). (**b**) Embrace the Stura experience (Source: International Workshop Design along the river. Turin, September–December 2014. Politecnico di Torino. Tutors: Ingaramo R., Voghera A., Claps P., Mutani G, Mazzotta A. Students: Alfarano E., Borra F., Pignataro A., Mattia Tempesta M)

research centres and social spaces, for establishing relationships between those who work in it and who want to relate to it (see Chap. 3 Ingaramo, design sketches).

4.4 Experimenting Design Tools and Methods

Starting from the river territories experiences, it was necessary to find a new approach to architectural design and town planning, in order to deal with the enhancement of the territory and landscape. This required the integration of projects for the transformation of nodes and larger areas, settlement models and service networks, relational spaces and green systems, both public and private, in accordance with the European Landscape Convention (2000).

With this in mind, our methodological issue – considered in the book Planning and Architecture (Ingaramo and Voghera 2012) – refers to a sustainable strategy which reconsiders the biophysical and social processes that have evolved as a result of development strategies, local regulations and planning, in terms of the opportunities they offer and the limits of physical and cultural resources employed in landscape enhancement projects, as already proposed in McHarg's ecological planning model (1969 and 1997).

The methodological proposal, the *integrated multi-scale projec*t, has no scale limits and can be applied to any area, be it regional, provincial or local, much like the architecture and design project. The method was tested in the Stura and Sangone workshops and during the Master's Degree course in Architecture and Environmental Territorial and Landscape Planning at Politecnico di Torino.

The method uses a tool to provide decision-making support and an operational technique. The Action Agreement is a participated planning instrument, helpful in supporting public and private decision-making consultation processes (see Chap. 2 Voghera). The transecting of sections is an operational technique used to correlate the various elements and scales of the project and to redesign territorial systems and complex landscapes (Ingaramo and Voghera 2012).

The integrated multi-scale project can help with decision-making by providing a continuous process linking the various scales of action and operating modes. It is also an attempt to overcome the hierarchical (and also temporal) vision of urban planning and architecture by immediately correlating structural and strategic decisions with the project for individual nodes, and the interaction of scales and thematic areas, traditionally dealt with using sectoral processes.

Natural, agricultural and urban areas play an equally important role in the layout of our territories. They have equivalent value in the integrated multi-scale project and help to identify comprehensive strategies for the comparison of the diversified tasks required to reconstruct the complex issues involved in sustainable landscape requalification.

The *Action Agreement* was created as a way of supporting strategic operational decisions in local projects, accomplished thanks to the consultation of public and private institutions and economic operators. This process helps overcome difficulties associated with bureaucratic decision-making, even when they arise during meetings between service providers to establish not only collective responsibilities but also those of an individual nature, regarding the implementation of established policies, dealing with the scenario of a Masterplan. The integrated multi-scale project is also related to the landscape enhancement project in line with the Cultural Heritage and Landscape Code (2004, subsequent amendments). It should begin with landscape planning, and it involves not only expertise but also pertinent responsibilities.

The *Action Agreement* could allow for negotiation between economic benefits and ownership interests. It could also be useful for the allocation of the local and economic resources necessary to guarantee implementation. Including it in the River Agreement could make the Masterplan visions and the local projects operational, mediating between different ownership interests.

Revisiting "Valley Sections" by Patrick Geddes (Geddes 1949) and New Urbanism transect planning experimentation (CNU 1993; Duany 2002; Duany et al. 2002; Brain and Duany 2005; Bohl and Plater-Zyberk 2006; Deakin and Allwinkle 2007), the method uses the transecting section technique. This uses three-dimensional sections to plan complex territorial and landscape systems, integrate various scales of action (the extensive area, boundaries between different land use areas, disputed territories, focus areas and architectural technological details) and create visions and projects for the enhancement of extensive and local areas to be discussed with process stakeholders (institutional, economic, public and private, social), in order to define overall strategies and individual transformation measures.

Transecting Sections are a tool for the representation and display of the physical morphology of the territory and the built area, of the relations that are determined by highlighting marginal empty spaces and the public and private spaces designed. They offer a vision for the construction of a system design rationale focused on morphology and uses.

This will reinforce the integrity and values of the various areas (built, unbuilt, natural, urban and rural), which can be considered as a continuum with different features, values and indexes of anthropisation/urbanisation.

There are interesting implications for town planning and architecture, as we can operate simultaneously in three-dimensional and two-dimensional modes, offering a supporting tool and verification of the hypothetical project. This operational method, which can also be implemented by non-experts, enables discussion with a view to establishing new organisational models on a town planning scale; new forms of living on an architectural scale (buildings, unbuilt spaces for which plans have already been drawn up, details such as fencing and furniture for public, semi-private and even private spaces); correlation between urban, natural and agricultural areas; and enhancement of identity and diversity.

References

Bohl C, Plater-Zyberk E (eds) (2006) Special issue: building community across the transect. Places 18(1)

Brain D, Duany A (2005) Regulating place: standards and the shaping of urban America. Routledge, New York, pp 293–332

Brower SN (2002) The sectors of the transect. J Urban Des 7(3):313–320

CNU (1993) Charter of the new urbanism. McGraw Hill, Ahwahnee, New York

CoE (2000) European landscape convention. Springer, Florence

CoE (2008) Recommendation CM/Rec(2008) of the Committee of Ministers to member states on the guidelines for the implementation of the European Landscape Convention. Adopted by the Committee of Ministers on 6 February 2008

Deakin M, Allwinkle S (2007) Urban regeneration and sustainable communities: the role networks, innovation and creativity in building successful partnerships. J Urban Tech 14(1):77–91

De Carlo G (2008) Questioni di Architettura e Urbanistica, Maggioli Editore, Santarcangelo di Romagna (RN)

Duany A (2002) Introduction to the special issue: the transect. J Urban Des 7(3):251–260

Duany A et al (2002) Special issue: the transect. J Urban Des 7(3):251–260

Geddes P (1949) Cities in evolution, in Outlook Tower Association and the Association for Planning and Regional Reconstruction. Williams & Norgate, London

Grahame Shane D (2011) Urban design since 1945. A global perspective. Wiley, New Jersey

Ingaramo R, Voghera A (2009a) Visione integrata e possibili evoluzioni. In: Concorso di progettazione. Contratto di fiume del bacino del torrente Sangone Masterplan del Piano d'Azione, Provincia di Torino. Universita gegli studi di Torino, Torino, pp 35–55

Ingaramo R, Voghera A (2009b) I Masterplan. In: Concorso di progettazione, Contratto di Fiume del bacino del torrente Sangone Masterplan del piano d'azione, Provincia di Torino. Universita gegli studi di Torino, Torino, pp 9–10

Ingaramo R, Voghera A (2010) Riverscape Valorisation project: Strategies and principles. The Sangone case study. In: Lehmann S, Al Waer H, Al-Qawasmi J (eds) Sustainable architecture and urban development, vol II. CSAAR Press, Amman, pp 115–131

Ingaramo R, Voghera A (2012) Planning and architecture. Searching for an approach. Alinea international, Florence

Magnaghi A (2014) La regola e il progetto. Un approccio bioregionalista alla pianificazione territoriale. Firenze University Press, Florence

Mazza L (2003) Trasformazioni del piano. Franco Angeli, Milan

Mcharg J (1969) Design with nature. Published for the American Museum of Natural History [by] the Natural History Press, Garden City/New York

Mcharg J (1997) Ecology and design. In: Thompson F, Steiner F (eds) Ecological design and planning. Wiley, New York

Voghera A (2011) Dopo la Convenzione Europea del Paesaggio. Politiche, piani e valutazioni/After the European landscape convention. Policies, plans and evaluation. Alinea Editrice, Florence

Part II
Case Studies

The second part reflects on case studies involving different river territories, environments and landscapes as strategic areas to develop equilibrium between human activities, hydrological flows and ecological values, affecting each other.

The river, a landscape mosaic, offers a reflection on design at different scales, focusing on the design process (De Rossi et al.), evaluations tools (Brunetta), governance methods (i.e. River Contracts; Berruti and Moccia) and specific landscape (Mellano) and urban topics and *τòποι*: urban fringe design and agri-urban models (Manzo), river parks and UNESCO sites (Sabaté Bel), greenbelts and ecological networks and urban microclimate (Mutani and Fiermonte).

Chapter 5
Territorial Governance in the River Districts: Methods and New Tools for Landscape Design Along the Po River

Grazia Brunetta

Abstract The chapter discusses innovations in the Italian river district policy with particular focus on the methods and practices that promote institutional cooperation and social responsibility. Keywords of this new perspective for the action along the rivers are *inclusiveness* and *multilevel governance* at the aim to implement the link between spatial policies and urban project. The paper shows the innovative vision in the Italian context that goes beyond the traditional sectorial approach in order to promote coordinated, synergic action of different policies regarding water, protection of the land, landscape and biodiversity. In this view, the recent experimentation promoted in the territory alongside the Po river configures an inseparable system of relationships between actions, objects and stakeholders linked to territory and landscape. The policies promoted by the Po River Basin Authority encompass not only sector planning actions but also uses and activities of the territory, involving territorial institutions, park authorities, other sector representatives and, in particular, the local community who become actor and must know, discuss, dialogue and share a complex urban project whose success is closely tied to joint participation in decisions. In the Italian context, the Po river corridor represents the natural backbone of Northern Italy, an economic resource for inhabitants and agriculture, a landscape that can become the main factor of attraction able to stimulate new sustainable rural uses and recreational tourism activities.

Keywords River basin planning • Po Valley Special Strategic Project • Strategic evaluation • Spatial planning

G. Brunetta (✉)
Interuniversity Department of Regional and Urban Studies and Planning,
Politecnico di Torino, Viale Mattioli 39, 10125 Turin, Italy
e-mail: grazia.brunetta@polito.it

R. Ingaramo, A. Voghera (eds.), *Topics and Methods for Urban and Landscape Design*, Urban and Landscape Perspectives 19,
DOI 10.1007/978-3-319-51535-9_5

5.1 Changes in River District Policies: From Sectoral Approaches to the Launch of an Inclusive and Integrated Institutional Governance Process

The governance of river basins has always been a central problem in our country, as proven by the frequent episodes of land instability and flooding, for which specific planning has been in place for more than 20 years. This planning has often had to cope with outcomes that have been anything but positive, with the local territorial projects and development actions. More than two decades after the launch of river basin planning in Italy,[1] the current situation highlights the complexity of the matter. Particularly an aspect has still not been fully solved, regarding the process of integration between the river basin plan and the territorial governance tools.

In the light of the river basin planning activities carried out during the early decades of implementation by the National Basic Authorities, some problems have emerged that are still ongoing and subject to specific experimentations and investigation in the current river basin governance phase. In short, we can highlight the following aspects in the centre of the current governance action in the river basins:

- A process of institutional cooperation among a group of institutional players – at central and local level – in order to "create system" among the different instruments and levels of governance of the areas influenced by the river basin
- A rationale of governance, founded the networking of levels of territorial governance and on the inclusion of the community (associations, enterprises, citizens, etc.) for the launch of the process of construction of the plan and of the project along the river
- A governance approach to the themes of land defence founded on the principles of responsibility and subsidiarity, in an attempt to settle disputes and diverging interests on the use of resources to strengthen the institutional cohesion in relation to the objectives of protection, enhancement and sustainable development of the territory

The need to strengthen the integration between planning/urban design and governance of water resources is an objective which has been emphasised by the strategies and international addresses that assign growing importance to the natural and landscape heritage and growing centrality to the "territories" and to the relative instances in policies for the conservation and enhancement of the environment and landscape (Council of Europe, UNESCO, IUCN, UN).

In Italy, there is an ongoing situation of separation of policies due to the breakdown of skills and the difficulty in integrating actions among them that belong to

[1] Despite the innovative capacity of Law 183/1989 having been sustained during the 1990s by important administrative reforms of the local autonomies, there have been considerable difficulties in implementing a planning process characterised by radically new content, methods and objectives.

different decision-making and operational systems. It is only recently (since 2007) that a new river basin governance phase has been launched, with the promotion, on a voluntary institutional basis, of inclusive planning processes that attempt to integrate or, at least, coordinate landscape and environmental requalification actions with projects to make the most of resources, settlements and local and regional economies.[2] It is a method of construction of the plan and the project that seems to be able to open up and welcome a new season of river basin governance, with a *governance* (rather than *government*) which could guarantee the operation and effectiveness, a project-building method that would appear to be suited to arranging the opportunities and resources that derive from the new European planning phase for 2014–2020 in the river "territories".

Overall, the current planning of the river basins seems to be heading towards the definition of a new phase of action, characterised by a strategic and inclusive rationale, aimed at implementing shared projects which concern not only the necessary works of mitigation of the hydrogeological risk but also of projects to make the most of territorial, environmental and ecological resources.

In this perspective, the process launched by the Po River District Authority (RDA) represents a pioneering experience at national level, with the elaboration of the "Operating Programme Scheme 2007–2013: A sustainable future for the Po River" (relating to the planning of the structural funds for the period), which tests and launches a new river basin planning approach, with the definition of "territorialising" objectives (agreed with the regional and provincial authorities), comprising not only water defence works but also nature-based projects, for landscape and environmental enhancement and fruition by tourism.

5.2 Innovation in the Territories of the Po River

Environment, territory and landscape represent areas of competence and actions, closely linked together but each with their own autonomy. They also represent the focus of attention of contemporary society which, in relation to the disruptive

[2]The institutional steps that have led to the construction of governance in the river basins can be summarised in the following framework of rules, agreements and conventions: (1989) Law 183 setting up the Basin Authorities – supra-regional organisation for land defence planning; (1992) Agenda 21 recognising the essential role played by comparison and communication between local institutions and bodies for sharing strategies for the sustainable management of territorial resources; (1992) Dublin International Conference on Water and the Environment, requiring the development of an integrated approach for the management of water resources; (1996) Law 662 introducing the negotiated programming for "regulation agreed among public bodies or between the pertinent public body and the public or private party or parties for the implementation of interventions, with reference to a single development aim"; (2000) Framework Directive on Waters 2000/60/CE; (2000) European Landscape Convention; and (2001) Strategic Environmental Assessment Directive 2001/42/CE.

dynamics of economic and social transformation of the period, is changing visions, values and expectations.

Starting with the presumption that effective results of enhancement placement in safety can be achieved by favouring integrated interventions in the sector of land defence and protection of water resources, the RDA has recently promoted, in order to implement the Hydrogeological Plan (Piano di Assetto Idrogeologico, PAI), approved in 2003, an inclusive process of cooperation with the local government authorities (provincial and municipal authorities). The process was launched in 2005 with the signing of the Protocol of Intentions by the Po River District Authority (RDA) and the 13 river provinces,[3] in order to accomplish defence and enhancement projects in the territory along the Po Valley (Mantua, 27 May). The Protocol defines the general objectives to be pursued and triggers a process of cooperation among the institutional players, based on principles of responsibility and subsidiarity. This process of governance, now rooted in these river territories, has become the engine for the activation of new design methods/territorial action, such as the River Contract.

This new multi-scalar and multi-sector district planning was launched specifically to strengthen the effectiveness of governance in river territories. It led to the creation, in 2007, of a negotiated planning document, a sort of "strategic agenda": the Po Valley Special Strategic Project (SSP) (CIPE resolution dated 21 December 2007 – approved in 2008) characterised by "territorialised" aims in terms of conservation, maintenance, requalification and fruition of the river territories.

This document marks the passage from a hydrographic basic governance rationale based on the exclusive prescriptive dimension of the sector, to a new version, defined also by planning addresses and guidelines for design, which, in addition to the essential actions to make the territories safe, with the design of water defence systems, also implements landscape and environmental re-naturalisation projects. This is the profoundly innovative vision of the intention to favour the coordinated and synergic activation of different policies, in keeping with the addresses of the recent European Directives on matters of waters, soil defence and biodiversity and of the European Landscape Convention.

The SSP represents the first step in this new path of action, in that it proposes the achievement of aims relating to the safety of the populations that live in the Po Valley, connecting them to those for the defence of the riversides, the improvement of the ecological network, the conservation of water resources and the fruition of environmental and historical-cultural resources. These integrated aims are the result of the multilevel governance process which, in addition to the promoters of the project – RDA and river provinces – has also involved regional authorities of Piedmont, Lombardy, Emilia-Romagna and Veneto and the river parks along the course of the Po and in the Po Delta.

In the governance model described above, the SSP is configured as a strategic agenda for achieving the qualifying and integrated aims that, when considered as a

[3] The Po provinces comprise the territories of Cuneo, Turin, Vercelli, Alessandria, Pavia, Lodi, Cremona, Mantua, Piacenza, Parma, Reggio Emilia, Ferrara and Rovigo. A total of about 500 municipalities are involved.

whole, represent an indivisible system of relations between actions, objects and subjects connected to the territory and the landscape. It is, in fact, obvious how a project relating to water resources is linked essentially to whatever takes place along the riversides, in the adjacent rural areas occupied by farms and settlements and in the town centres crossed or adjacent, directly involving a broad spectrum of human activities that take place there. This perspective of action is also in line with recent Italian legislation on landscapes, which reiterates the absolute priority of combining defence with enhancement, networking the territory's environmental, natural and cultural resources to offer them to an adequate activity of fruition by the population. Consequently, the policies promoted by the RDA come into play, as well as the actions relating to sectoral planning, interests, hopes and demands of the local communities settled along the riversides. These demands are expressed in an extended and participated form of construction of the territorial planning and programming process, involving the territorial entities, the park entities and all the other entities appointed to manage specific and sectoral aspects. Different subjects go from being spectators, competitors and users to players – each with their own responsibilities – to know, communicate and share a complex project, the success of which becomes closely linked to participation in the decisions, their planning and implementation.

With this in mind, the SSP touches on projects for the defence and enhancement of the territories throughout the whole of the Po Plain; the river course becomes the main natural structure of Northern Italy, a great resource also at economic level for the cities, settlements and agriculture. A landscape, which once requalified and equipped, can become extremely attractive not only for the local populations but a place where new sustainable rural, artisanal, commercial and recreational tourism activities can be developed. It is a large-scale project of national scope to be implemented, becoming an experimental model of application of the most recent European Directives and national legislation on matters of waters, with the aim of promoting a better environmental, social and economic sustainability in the planning of the hydrographic districts.

5.3 The Launched Innovation Process: The Methodology of Assessment for the Construction of Governance in the Po River Territories

In this particular institutional context of reform that parallels to the process of construction of the Po Valley SSP and the design and experimentation of the Strategic Environmental Assessment (SEA) methodology[4] which, in these particu-

[4] The Strategic Environmental Assessment methodology is the result of the research project between the Po Basic Authority and Turin Polytechnic Corep, "Pilot Project for the application of the strategic environmental assessment to basin planning implementation in the territorial plans" (2007–2008), carried out by a multidisciplinary study group and coordinated by Grazia Brunetta and Attilia Peano. Work group: Grazia Brunetta, Patrizia Lombardi, Attilia Peano, Agata Spaziante

lar conditions, have been able to perform the function of technical accompaniment of the political decision-making process relating to the prefiguration of plans and projects for re-naturalisation, enhancement and development along the river.

The rationale of the planned assessment approach lies initially in the fact that it takes place parallel to the process of governance launched between the RDA and the Po river provinces, gaining importance as a new cognitive "tool" for institutional learning, to build a series of shared criteria to help define priorities and options for defence and enhancement in the Po river territories.

The aim of the assessment methodology planning is to strengthen the integration between the plans and policies that coagulate around the river, and it acts as a strategy to reinforce the governance of the Po river district in the local territories. In this sense, the assessment method has been designed for configuration as technical support for the pursuit of the planning process in the river territories.

Two basic conditions were assumed as the main tenets for the design of the SEA methodology.

The first concerns the type of object to be assessed: not a single plan or programme of the Po River District Authority but the river basin planning, that is, the comprehensive view launched by the plans and programmes (within the competence of the Po River District Authority) that impact basin territorial, environmental and landscape resources.

The Po River District Plan is a cross-scale, multi-sector planning process, launched following the approval of the "basin plan" (2003) and currently being revised towards a strategic approach perspective. The SEA methodology applied to river basin planning has been configured in such a way as to facilitate the planning process in the territories of the basin (from the river to local area). The Po Valley Special Strategic Project has led to a first agreement with the river provinces for the prefiguration of a series of objectives and actions to outline in the local projects for the enhancement of territorial, environmental and landscape resources.

The second concerns the need to manage the complex processes of governance in the Po river territories which, regarding the country's broad, varied and complex institutional and social reality, refers back to the consideration of numerous different demands for the use of resources in progress. With this in mind, the assessment and planning method adopted takes on particular importance, also with regard to the involvement of the community (municipal and provincial authorities, stakeholders and citizens) affected by decisions relating to environmental enhancement and territorial development. This makes it necessary to manage the conflicts of use of the land by implementing processes of cohesion among subjects and levels of governance, as an essential condition for improving the effectiveness of policies and projects in the river territories.

These two conditions become the assumptions for the assessment methodology experimented in planning river districts, as applied to the Po Valley Special Strategic Project (SSP), characterised by a strongly integrated territorial perspective that

(Turin Polytechnic), Luigi Bobbio, Patrizia Saroglia, Fiorenzo Martini (Turin University) and Paolo Urbani (Pescara University). See Brunetta and Peano (2010).

embraces not only conservation of water resource quantity and quality but also protection of the river banks, improvement of the safety of local populations, strengthening of the ecological network and improved fruition of environmental and historical-cultural resources and river tourism.

The main ethos of the evaluation approach adopted for Po river district planning is that it is carried out in parallel with this ongoing decisional process, with the aim of contributing to the construction of a reference framework (regarding the selection of priority criteria) for the prior definition of shared intervention options and goals.

5.3.1 The Evaluation Process

The SSP is configured as a *strategic plan* that proposes a new vision of the Po river district, based on the achievement of three *main objectives*:

- 'territorialisation' of the Po river Basin;
- planning and design the Po river basin as a system;
- integration between territorial policies and urban projects along the river.

The achievement of these aims is articulated into four *areas of action*: (i) improvement of hydraulic safety, increase in the lamination capacity of the river banks and morphological reconstruction of the flood plain; (ii) conservation of the ecological integrity of the river banks and water resources of the Po river; (iii) the system of fruition and cultural and tourist facilities; and (iv) the system of governance and of intangible networks for knowledge, training and participation.[5] These four areas of action of the project – enhancement of hydraulic safety conditions, conservation of the ecological integrity of the territory alongside the Po river and of water resources, system of fruition and cultural and tourist facilities and governance – configure an inseparable system of relationships between actions, objects and stakeholders linked to the territory and landscape. The policies promoted by the Po River District Authority encompass not only sector planning actions but also uses and activities of the territory, involving territorial institutions, park authorities,

[5] The first objective, the flood and landslide risk mitigation plan for the Po river district (Piano per l'Assetto Idrogeologico – PAI), represents the main planning tool to identify prevention and mitigation actions regarding hydrogeological risks (complying with the EU Floods Directive). The second objective deals with actions finalised at the strengthening and expansion of the ecological network and at a better-integrated management of water resources, including environmental uses. Along the River Po, many CISs and SPZs belonging to the European Natura 2000 Network have been identified. With regard to these aspects, the Strategic Special Project aims to promote the completion and coordination of management plans of Natura 2000 areas, maintaining the vision of the whole river system. The third objective focuses on the potential natural and cultural heritage value of the river territories and brings forth actions to improve territorial appeal and sustainable economic activities and tourism. The fourth objective crosses all the other three and deals with the strengthening of the governance system. Jurisdictional fragmentation between different institutional levels and complexity of communication between different actors make it necessary to develop cooperative models to achieve effective results.

other sector representatives and, in particular, the local populations who become actors and must know, discuss, communicate and share a complex project, the success of which is closely tied to joint participation in decisions and their application. The SSP is a major national project that embraces the entire Po Plain, four regions and all the river provinces, close on 500 municipalities; the river corridor represents the natural backbone of Northern Italy, a major economic resource for inhabitants and agriculture, a landscape that can become a major factor of attraction capable of inspiring new sustainable rural, craft, commercial and recreational tourism activities.

The structure of the Strategic Environmental Assessment process has been defined according to areas of action and key issues of the SSP:

- Water resources
- Soil
- Biodiversity
- Landscape and environmental assets
- Hydrogeological risk

These five key issues become the environmental category clusters of the assessment process. The structuring of the process takes into account the strong interrelations between the environmental category clusters and decision-making and organises the information according to *two levels of assessment*.

The first level, two-dimensional, concerns the identification of the relationships between the *areas of action clusters* (and the actions within each area) and the five *environmental category clusters*. This level of analysis highlights only the hierarchical relationships within each area of action and those within the same cluster, without taking into account the multiple, complex relationships that exist, reciprocally and generally, between the various areas of action and environmental category clusters.

The second level of analysis permits the structuring of assessment according to a multidimensional logic. The decision has been made to adopt the multi-criteria technique of the analytic network process (ANP)[6] which makes it possible to move from a two-dimensional to a multidimensional type of analysis and assessment, capable of highlighting the complex links between the areas of action of the SSP. In fact, the ANP is the first mathematical approach that makes it possible to systematically deal with all kinds of dependencies and feedback among elements. It requires the identification of a network of clusters and nodes, as well as a pairwise comparison to establish relations within the network elements.

The model has made it possible to achieve a first ranking of preferences, as the outcome of the comparison of pairs among all the possibilities that made up the

[6] The ANP is a technique which allows the passage from a two-dimensional to a multidimensional assessment, in that it is distinguished as an evolution of the analytic hierarchy process – AHP (Saaty 2001) hierarchical model. In this sense, it envisages the structuring of decision-making in the form of a network and the same method of comparison of elements using comparison by pairs. See Lombardi (2005); Lombardi et al. (2014).

decision-making process.[7] The decision-making model planned was experimented with two focus groups, organised in Parma with the technicians of the RDA and the members of the work team, with the aim of proposing a first order of lines of action and subjects of intervention for inclusion in the plan and to share with the institutional actors involved in the governance process. The ANP model has become a technical tool to support the decision and has been considered as an open model for the representation of the deliberative elements, in that those involved in making the decision have the possibility to change the structure of the model, varying the links between the nodes of the system of relations.

The ANP model was tested during two workshops with Po River District Authority technicians (in Parma on 21 May 2008), including it in the process of preparation of the SSP as a basis for prioritising the areas of action and issues.

5.4 Assessment of Results: Designing the Landscape Project for a Sustainable Future Along the Po River

In the technical experimentation carried out, the assessment methodology returned a ranking, according to the assessment criteria, that places major weight on the node of "enhancing the river area's natural and cultural heritage", which embraces polices for the safety of the land, water management, conservation of biodiversity, land use, fruition and enhancement of resources. The final ranking of the environmental category clusters stressed the priority of the "landscape, environment and cultural heritage and rural spaces" cluster, inserting the SSP as an integrated landscape policy within the framework of the European Landscape Convention. The assessment method reveals the in-depth innovation introduced by the SSP compared with traditional sector-type river basin policies, which have often proved to be of limited effect, being based exclusively on sector-level actions and environmental and territorial constraints. Recouping of landscape policies in order to develop actions addressing environmental conservation and requalification, safety of the populations and activities and suitable valorisation of fruition in order to promote a balanced and sustainable development of the territory represents the new frontier proposed by international guidelines.

The framework outlined by the assessment process configures a complex project, now in progress, sustained by a system of governance based on scientific knowledge, technical training and social participation. Therefore, the Po River Basin planning process, started with the aid of the SEA methodology, outlines a planning action perspective to be completely implemented and which is undeniably difficult

[7]The ANP model translates the subjective judgments of comparisons by pairs, expressed by the participants in the focus group, on a 9-point scale, allowing the definition of a ranking of the decisional elements, reached through three steps: (1) elaboration of matrices of comparison by pairs, (2) aggregation of the results and (3) sensitivity analysis. The elaboration of the results is made possible by a specific software.

to apply in view of its complexity, but which certainly promises virtuous results in terms of quality and effectiveness.

According to the assessment methodology experimented, the Po River Basin planning should move towards an innovative planning process. This should be inclusive and oriented to implement shared plans and projects regarding necessary works to mitigate risk and actions aimed at conservation and the environmental restoration included in the framework of sustainable development. Environment, territory and landscape are fields of strongly related study, skills and action, each with their own relative autonomy. These areas address the major international strategies that will guide territorial policy in the third millennium: *conservation of biodiversity*, for the life of the planet and strongly linked to health and economy; *protection of landscape* as an ecological, cultural, social and economic resource; and *sustainable development* as a guiding principle of the most appropriate relationship between human activities and the global environment.

The landscape embodies values that have emerged in contemporary society and which involve collective and intergenerational interests, establishing new rights and new duties, tied to ongoing in-depth changes: transit to the post-industrial age, economic and social globalisation, the advent of the digital era and the role of communication and image. Which prospects may unfold for landscape planning? What role can it play in redefining the relationship between planning and society? The landascape implies issues of ecology, of structured permanence and of perception, while also preponderantly addressing matters of territorial governance, such as land use and safety, management of water, consumption of territory, depopulation, infrastructure sprawl, the conditions of rural areas and the rapport between city-nature-agriculture and the local economy. Questions and policies that involve the contemporary social project and which impose the need for innovative technical, administrative, regulatory, planning instruments capable of accompanying transformation processes rather than being imposed are open to dialogue, confrontation and incentives. Landscape and territory are, therefore, inextricably linked in order to construct a prospect of "sustainable" progress of society. A path is fraught with difficulties as it implies overcoming the fragmentation and sectoral nature of laws, rules, financing and organisation of the public administration at all levels and also encouraging private operators to take a vision of the future impacted by the effects, also at economic level, of improved management of the territory and landscape.

5.5 Final Remarks

The cognitive framework that has emerged from the assessment process has made it possible to support the configuration of a heterogeneous project of river basin governance, based on a broad and articulate territorial governance system in terms of the actors and institutional contexts involved. The assessment method, included in the renewed logic of the action that the Po River District Authority had launched at that time, has represented the technical contribution to the launch of the new governance perspective that has been consolidated in recent years along the river territories. The new River Contract tools, which are currently undergoing experimentation to promote responsibilities, knowledge, cooperation and subsidiarity, have set themselves the aim of recognising and integrating that framework of different and often opposing values that emerge along the river. This is definitely an open and promising perspective, but it is undoubtedly harder to put fully into practice.

Despite the partial nature of this first experimentation, the assessment method shows just how useful it is, in that it is structured as a process and an aid to the construction of the action to configure the aims and priorities of the enhancement projects to promote along the Po river. In particular, thanks to the application of the network model, it has been possible to systematically take into account the multiple interactions existing between environmental conditions, anthropological activities and the need to improve the hydrogeological conditions of the river territories. This has brought us to a realistic representation of the decision-making issue, which is still a clear element of strength of the assessment/planning/design approach. The assessment method has incorporated and represented the variability and the turbulence of the design and implementation process of the Po Valley SSP which, as previously mentioned, is the strategic agenda for action and for the accomplishment of projects.

The way of life in these areas has changed. There are new relationships, new places and new landscapes which require innovative technical, administrative, regulatory and design tools, more of an accompaniment of the transformation processes than an imposition, open to communication, comparison and incentives. The assessment method is a cognitive process to accompany the process of elaboration of the choices of the plan, which allows the recomposition and constant update of the framework of action, starting with the definition of certain shared invariants (the criteria and environmental categories that emerged from the ANP). From this point of view, another evident strength of the process is the contra-intuitive result highlighted by the assessment, for the launch of landscape policies for the Po Valley.

The assessment method, as recognised by the technicians involved in the process, has performed an effective institutional role in learning to launch and affirm the logic and the perspectives of action of the new course of planning of the Po River Basin.

The results of the experimentation reveal the potential utility of the assessment action, as the technical process for organising the construction of shared plans and projects for enhancement and development along the river. A complex and, consequently, "unstable" path in terms of the results will be outlined, via an approach which intends to incorporate the actions of the plan with those of the design. With this in mind, as shown by the experience described, the assessment method has launched the negotiation and design procedure along the river, innovating the capacities for definition and design skills along the river.

The case shows the essential integration between assessment/planning/design – a necessary condition to achieve substantial improvements within the framework of action and design using the assessment method, as we are reminded by eminent researchers who have played a pioneering role in the field (Faludi and Voogd 1985;

Khakee 1998; Guba and Lincoln 1989; Lichfield et al. 1975; Lichfield 1996). In this logic, the assessment seems much more effective and useful if considered as a cognitive technique in support of the decision-making cycle/action, or the process to be carried out parallel to the elaboration of the plan and design, to accompany the construction of sustainability scenarios. It is a question of taking the assessment dimension as a strategic framework within which to proceed with a progressive composition of interrelations among the resources involved in the decision-making and design process. Within the scope of the decision between public subjects, the hierarchies among *values and criteria relating to choice* and their relative weight are not just figures, but are the fruit of agreements between different governance actors. This aspect requires representation and comparison, in the construction of decisions relating to the plan and design, of the different points of view which appear, from this perspective, all legitimate and which cannot, therefore, be treated according to a perfectly linear logic nor coordinated according to a sequential hierarchy established in advance. The assessment in planning is built *in doing* and is updated ex post *incrementally*, seeking an answer to those questions which open up during the action. This *strategic* assessment approach is focused on the construction of a shared technical-cognitive framework, via the introduction of meta-criteria aimed at gradually recomposing the positive effects generated during the decision-making/action process, to promote new possible forms of agreement. Assessment, therefore, takes on the meaning of a technical argumentation within the drawing up of the plan and the design, not only as a discursive structure for deciding but as a process of institutional learning and progressive accumulation of knowledge to be used in the action and organisation of design (Alexander 2006; Brunetta 2015; Lichfield et al. 1998; Miller and Patassini 2005). In this sense, the potential added value of the assessment method emerges, if interpreted as a possible (but not the only) innovation in the river territory governance approaches, to incorporate environmental matters into the process of construction of the plan, in an attempt to raise the level of acceptability of decisions and designs. From this perspective, the role of institutional design that the assessment method can perform in sustaining innovation in the process of design of territorial improvement and requalification actions also seems clear (Alexander 2005, 2009). The gradual learning function that the assessment can perform in the plan process is manifested with the introduction of new knowledge and technical ability; its capacity should be wide enough and dynamic enough to be able to support the decision-making/planning/design process, especially in those territories – the river basins – that are particularly sensitive, due to the high level of conflict between the interests involved.

References

Alexander ER (2005) Implementing norms in practice. The institutional design of evaluation. In: Miller D, Patassini D (eds) Beyond benefit cost analysis. Ashgate, Aldershot, pp 295–310

Alexander ER (2006) Evaluation and rationalities: reasoning with values in planning. In: Alexander ER (ed) Evaluation in planning. Ashgate, Aldershot, pp 39–52

Alexander ER (2009) Dilemmas in evaluation planning, or back to basics: what is planning for? Plann Theor Pract 10(2):233–244

Brunetta G (2015) Smart evaluation in regional development. territorial scenarios in Trentino, Italy. Ashgate, Aldershot

Brunetta G, Peano A (eds) (2010) La pianificazione di bacino verso politiche integrate. Urbanistica 143:7–31

Faludi A, Voogd H (eds) (1985) Evaluation of complex policy problems. Delftsche Uitgevers Maatschappij, Delft

Guba EG, Lincoln YS (1989) Fourth generation evaluation. Sage, Thousand Oaks

Khakee A (1998) Evaluation and planning: inseparable concepts. Town Plann Rev 69(4):359–374

Lichfield N (1996) Community impact evaluation. University College Press, London

Lichfield N, Kettle P, Withbread M (1975) Evaluation in planning process. Pergamon Press, Oxford

Lichfield N, Barbanente A, Borri D, Khakee A, Prat A (eds) (1998) Evaluation in planning. Kluwer Academic Publishers, Dordrecht

Lombardi P (2005) L'Analytic Hierarchy Process e i suoi sviluppi. In: Roscelli R (ed) Misurare nell'incertezza. Celid, Torino, pp 41–60

Lombardi P, Giordano S, Brunetta G, Caldarice O (2014) Applying the analytical network process to identify new retail scenarios in the Province of Trento (Italy). Int J Analytic Hierarchy Process 6(2):145–158

Miller D, Patassini D (2005) Introduction – accounting for non-market values in planning evaluation. In: Miller D, Patassini D (eds) Beyond benefit cost analysis. Ashgate, Aldershot, pp 1–11

Saaty T (2001) The analytic network process. RWS Publications, Pittsburgh

Chapter 6
Planning Knowledge and Process for Strategies of Participatory River Contracts

Gilda Berruti and Francesco Domenico Moccia

Abstract In the river contract process, an ever more widespread participatory policy in Europe, the first step never enough considered is the construction of an appropriate knowledge and a streamlined planning process. The main challenge to face is the collection of multidisciplinary requirements coming from the different domains that can contribute to the analysis of the river basin, thus overcoming shortfalls and problems that can arise in the process and stimulating change management scenarios. In this complex setting, the planner's perspective seems to be the most appropriate to call sciences to offer their contributions for the aims of the program because they must fit in a multi-actorial decision process aimed to pragmatic actions. According to this perspective, some other questions are added to the consolidated knowledge on the river basin: from the study of the hydrographic network and reports on soil quality to the analysis of urban and social systems, a focus on connections and interactions, and a study on economic activities. This knowledge comes from heterogeneous sources, such as information officially produced by government bodies, found in the scientific literature, or resulting from oral stories and interviews.

From the sharing of this framework by public and private actors, the river contract starts and develops with an action plan that participants commit to carry out signing an agreement. With exercises in a wide list of cases, many recovery potentials are identified and strategic redevelopment paths indicated.

Keywords Strategic planning • Participation • Landscape restoration

---

G. Berruti (✉) • F.D. Moccia
Department of Architecture, Federico II University of Naples,
via Forno Vecchio n. 36, 80134 Napoli, Italy
e-mail: gberruti@unina.it; fdmoccia@unina.it

R. Ingaramo, A. Voghera (eds.), *Topics and Methods for Urban and Landscape Design*, Urban and Landscape Perspectives 19,
DOI 10.1007/978-3-319-51535-9_6

6.1 Introduction

In the latest years, we can observe a shift toward riverfront development, from industrial areas and urban backs to public spaces belonging to the connective and recreational fabric of the city. This shift can be observed in American cities in the last decades of the twentieth century, more recently in Europe.

Engineered interventions of the past carried out to control rivers in relation to the rise of the so-called bacteriological city (Gandy 2004) and the consequent hygienist emphasis are being overcome by a wider acknowledgement of the complexity of river ecosystems. Distinctive urban technological landscapes, in which water channeling infrastructure prevails and rivers are concealed, contemporarily working to control water pollution and provide new spaces for car movement and parking, as effects of the "impulse to transform nature in the service of a new society" (*ib.*, 368), are not the order of the day anymore.

The objective is to bounce rivers forward to a state not so far from the natural one, depending on the context's situation, according to an evolutionary resilience approach (Davoudi 2012) that implies an idea of evolution, which well attains to the complexity of cities as socio-ecological ecosystems.

The urgency to take care of rivers not only as physical components of natural systems but as strategic elements for the quality of life of global population is ratified by the EU Water Framework Directive (2000/60/EC) establishing a track for community action in the field of water policy. The directive pursues sustainability in its ecological economic and social dimensions and provides a basis for the development of strategies toward a further integration of policy areas.

According to the environmental objectives of the directive, member states shall protect, enhance, and restore all bodies of surface water with the aim of achieving "good water status" by 2016. To reach this objective, the directive requires river basin management plans to be developed for each river basin district. This condition is still difficult to achieve for many Italian rivers that are in a dreadful state, since for decades they took in the wastes coming from the development of the surrounding areas, including industrialization and agricultural processes.

Nevertheless, water pollution is not the only problem to solve. The objective of EU Directive 2007/60/EC on the assessment and the management of flood risks is the reduction of adverse consequences of floods for the human health, environment, cultural heritage, and economic activity. According to this directive, in each river basin district, the flood risks and need for further action should be assessed. Flood risk management plans including tailored solutions according to the needs and priorities of the investigated areas become necessary at the level of the river basin district and must be carried out in coordination with or integrated into river basin management plans (EU Directive 2000/60/EC).

River floods, increasing with the warming in the XXI century, are reported among freshwater-related risks of climate change for global population by the

Intergovernmental Panel on Climate Change (IPCC 2014) focusing on risks and vulnerabilities of contemporary cities. Unintended consequences of environmental actions are also pointed out: "diverting a river course or building an embankment to protect new development may prevent flooding in one location, but may cause or increase flooding somewhere else" (IPCC 2014, 547).

The awareness of the role that river corridors have with respect to the health of urban ecosystems is rapidly growing together with the consciousness of the impacts of human activities on the river flow and its variability in time. The river system can be described as a connection of the patches of the landscape mosaic, "a *river continuum*, in which water flow, organic matter, fish populations, and many other factors change somewhat gradually from headwaters to mouth […] in most areas the matrix, the floodplain, and the river itself are highly patchy, with relatively distinct boundaries" (Forman 1995, 209–210).

River corridors are in dynamic equilibrium. Four dynamic processes take place in river landscapes: hydrologic flows, particle flows, animal activities, and human activities (Forman 1995, 216), each one influencing the others. These processes affect very different territories, going from the urban context to the countryside, and it is essential to find the way to properly react according to circumstances. It is necessary to reason on a regional scale, looking at river basins and investigating the different components of the involved socio-ecological ecosystems. From this point of view, river contracts may be useful tools to restore both ecosystems and the sense of community at local and regional scale.

This contribution focuses on river contracts as water management instruments that require an appropriate multidisciplinary knowledge and streamlined planning process to be constructed and launched. First, the meaning of river contracts and their place in Italian legislation are outlined. Then, the construction of an appropriate method that may reveal useful to the river contract process is investigated, from the collection of data and facts that constitute the cognitive framework of the river basin to the arrangement of the relative action plan and the involvement of all the local relevant actors. Simulations of different stages of river contract processes carried out by the authors during urban planning courses at the Urban, Landscape, and Environmental Planning Program,[1] Federico II University of Naples, are adopted as examples.

[1] In particular, examples come from the third studio on "The Regional and Landscape Plan," instructor: Francesco Domenico Moccia, Fall semester 2013 and 2014, Urban Landscape and Environmental Planning Program, and several master thesis on the same themes, together with the course "The image of river landscapes. Survey on the management of riverfront territories," instructor: Gilda Berruti, Fall semester 2014. Examples concern: Sarno River and its tributaries Cavaiola and Solofrana, Tanagro, Alento, Irno, Sebeto, Calore Irpino, and Sabato in the Campania region of Italy, Tordino River in the Abruzzo region, Fibreno and Aniene Rivers in the Lazio region, a section of the Po River in the Emilia-Romagna region, Biferno and Basento Rivers in the Molise region, and Virginio stream in the Toscana region.

6.2 The River Contract as Voluntary Negotiation Tool

The river contract is a governance institutional process for water management at local level and aims to involve all the main actors of the area in the identification of integrated actions of river environmental policy intended to foster the river, considered as a value resource for the recovery and development of territories. In so doing, it is possible to overcome the sectorial logics often characterizing the way of tackling environmental issues by the public sector.

The river contract can be useful to solve problems mainly pertaining to water quality and the river health condition; security of river valley, from hydrogeological risk to landslides; and environmentally aware fruition of river banks. In Italy, the river contract is not primarily instituted by law (Voghera and Avidano 2012); it is "more of an experience implemented and developed in the last ten years and constantly consolidated both methodologically and operationally" (Voghera 2015, 352).

The river contract is a voluntary agreement among several public and private subjects that is constructed through negotiations and assumes that each of them will take his responsibility to carry out the undertaken commitment.[2] As a matter of fact, it is the contract as the tool that could assure the implementation of projects defining the future of river territories; its role is to design clear and shared rules of the game with the involvement of local stakeholders. Some legal norms and public programs come after to register what has already happened in practice.

From a formal point of view, in Italian legislation EU Water Directive was enacted in the Italian Legislative Decree n.152/2006 that focuses on rules in environmental field and decides on the integration between water management issues and regional planning. Decree n.152/2006 introduces the hydrographic district,[3] defined by the unification of adjacent hydrographical basins, and institutes the Regional Water Protection Plan, aiming to establish rules for the protection of water courses interacting with regional and development policies. In this framework, there is no mention of river contracts.

Some regional governments, such as Piemonte, Lazio, Lombardia, and Marche, enacted river contract guidelines and allocated financial appropriations to this effect in their budgets. Only recently, in December 2015, the Italian Law n. 221 on measures to promote green economy and the control of the excessive use of natural resources was approved that recognizes river contracts and establishes their objectives (art. 59): river contracts contribute to define and carry out planning district instruments as voluntary instruments of strategic planning, pursuing the protection, the correct management of water resources, and the valorization of river territories,

[2] In France, where the *Contracts de rivière* were experimented for the first time in the 1980s, the turning point between first-generation and second-generation river contracts, respectively, oriented to a sectorial management of the water course and an integrated one, is defined by the law n. 92 of January, 3, 1992 (Cfr. Bobbio and Saroglia 2008).

[3] Italy has eight river basin districts: Eastern Alps, Po Basin, Northern Apennines, Serchio, Middle Apennines, Southern Apennines, Sardinia, and Sicily.

together with the safeguard against hydraulic risk, and being a factor in the local development of hydrographic basin's areas.

The law comes after many efforts done by the National Table of river contracts[4] in order to prepare the National Charter of river contract (2010 first edition) and the quality requirements of river contracts (March 2015) in a work group coordinated by the Ministry of Environment, Sea, and Land Protection, together with ISPRA. In the aforementioned requirements, river contracts are identified as operational instruments able to provide operative and checkable outcomes in the short/average period finalized to solve concrete problems of a specific area coordinating already existing plans and programs and the local interests.[5] Important elements to develop river contracts are the start of participatory processes, to identify resources and problems of a territory and to define appropriate actions and relative responsibilities, to join together institutions and local governments responsible of water and rivers, and the coherence of river contracts with territorial social and administrative contexts and the current rules.

The work group will continue its activities to identify useful indicators to monitor river contracts for their quality and progress. The creation of a National Observatory on River Contracts at ISPRA was also established with the aim to provide an up-to-date framework of river contracts in Italy and to assess their quality and results.

As far as organization requirements are concerned, the different stages of a river contract are:

- Sharing a document containing preliminary objectives, organization bodies and the decision process, and methodology to adopt. Signing the document marks the start of the river contract process.
- Developing an integrated knowledge framework about the river basin.
- Conceiving a strategic scenario able to combine large area planning objectives with local development policies of the analyzed territory.
- Defining an action plan with a clear time line (three years maximum) to be updated at the deadline.

As it is possible to observe, the role of stakeholders is crucial in a river contract process. According to Bobbio and Saroglia (2008), participants in a river contract can be distinguished in three groups. There are actors looking "inside the river," mainly experts, interested in the safeguard of water quality and quantity, and actors looking "around the river," politicians and economic subjects, interested in local development and river users, individual, or associations enjoying living river

[4]The National River Contract Table was born in 2007 as an Italian Agenda 21 working group, with the aim to consolidate a community able to exchange experiences and promote river contracts in Italy. Regions, provinces, towns, entrepreneurs, and organizations working to launch strategies to recover rivers, lakes and coasts with inclusive processes contribute to the table.

[5]The document dwells on the concept that river contracts do not constitute a new level of planning.

Table 6.1 River contract process

Sharing preliminary objectives	Integrated knowledge framework	Strategic process	Action plan
Organization bodies	*Issues:*	Initial agreement	*Issues:*
Decision process	Hydrographic network	Mission statement	Ecosystem functioning
Methodology	Soil quality	SWOT analysis	Responses to risks
	Urban and social system	Strategy development	Riverfront fruition
	Connections and interactions	Strategic territorial scenario	Management of river landscapes
	Economic activities		Stakeholders
	Problems and potentials		Time line
			Implementation process
			Resources

territories that do not seem to look at the river at all or look inside and around the river at the same time.

Different views imply different effects on the territory. The first group's view involves an attention to the river on its own in its ecological features, to be protected from human activities and to make not dangerous for the surrounding environment. Conversely, actors looking "around the river" consider it a resource for bordering territories and the center of local development policies. Both these visions are incomplete, and a balance among them is necessary that river contracts should be able to provide, overcoming power asymmetries among partners (Bobbio and Saroglia 2008).

In addition to this stakeholder analysis, practicing many processes, as those listed forward, especially in the starting step, entails to be aware that the first move to consider a river contract comes from risk, a pervasive issue in public opinion that affects both people and leaders. This pulse generally has an impact on the complexity of public action for three main factors: the number of public organizations that share the responsibility and difficulties of cooperation when going over the routine control of regulations, the collection of resources generally segmented among many governing bodies, and conflicts arising when an environmental problem is touched and the people's health is involved. The cooperative process of river contracts appears as a way to overcome these difficulties because it assures a forum where people and public officials may listen to each other and public organization can cooperate from the start, planning together.

The expected product, at the end of the process, is a framework of objectives about river landscape recovery and development, organized in lines of action and going with a strategic design vision, the so-called action plan that partners have to share and to sign, committing to put it into practice (Table 6.1).

6.3 Toward a Method: The Construction of a Knowledge Framework for River Contracts

In a participatory process, the main question for planners is: How can we contribute? What is our role, avoiding impinging on the interests of the involved parts and supporting the decision process? These questions are central in planning practice as well as in teaching. That is why the first discussion in our planning studio focused on how to define limits and contents of planners' contribution.

In finding such position, we must consider a setting where initiatives aiming to develop river contracts were few. In a situation of low reputation of planning and hard attitude to social cooperation, the first identified role of the planner seemed to be that of initiator. As some students were employees of municipalities or other local administrations, others were involved in planning practices, they identified potential stakeholders among people they trust and thought interested and presented advantages of the process. To make clear what they could expect, three points were essential: a survey of the river basin to show problems and potential solutions; an outline of the planning process, its steps, and the organization needed; and potential strategies and projects. All was intended in a work of stimulus to find and activate stakeholders in the first stage of the planning process whose development should be determined by them in the practice. Certainly this was something that we could not create inside the university. This is the reason why much of the work focused on the knowledge construction.

As a matter of fact, the construction of a comprehensive knowledge is the first step to take in the direction of a successful river contract process. The main challenge to face is the collection of multidisciplinary requirements coming from the different domains of knowledge that can contribute to the analysis of the river basin, thus overcoming shortfalls and problems that can arise in the process and stimulating change management scenarios. In this complex setting, the planner perspective seems to be the most appropriate to call sciences to offer their contribute for the aims of the program. According to this perspective, some other questions are added to the consolidated knowledge on the river basin.

The proposed framework involves different branches of knowledge. First of all, it includes an in-depth study on the hydrographic network that aims to identify the river and its tributaries and the changes arisen in the water courses and human interventions on the river banks, the investigation into the quantity and quality of water and the detected sources of pollution, and the presence of hydrogeological and hydraulic risk in the hydrological basin (Fig. 6.1). Here, the idea that the hydrographic basin is the unit of analysis should be stressed, and its identification is the first step of the knowledge process. This operation could be somewhat difficult in urban settings whereon ground water dynamics were strongly altered. Second, it reports on soil quality, describing soil and land uses and the involved geomorphological processes. Third, it analyzes urban and social systems, pointing out existing hierarchies among small towns and higher-level centers, demographic dynamics, physical interactions with rivers, and how the river and its banks are used (Fig. 6.2).

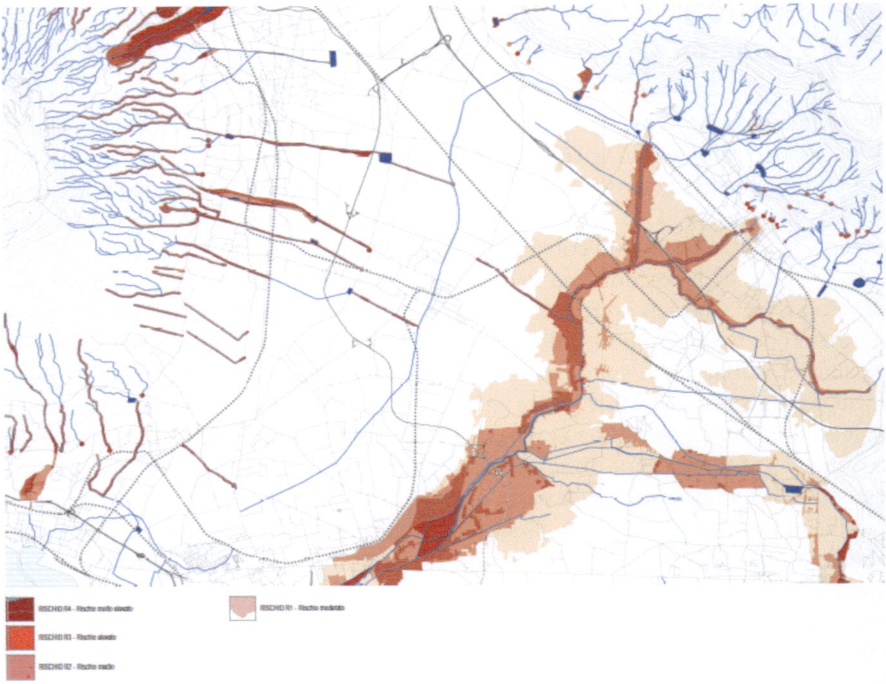

Fig. 6.1 Hydraulic risk in the Sarno Plain (Source: Laboratorio Piano Territoriale e Paesaggistico, students: Sammarco F. S., Sepe C., Vinaccia D.)

Fourth, the knowledge framework focuses on connections and interactions, drawing attention to interdependence among urban centers and between urban centers and their hinterlands; accessibility and transport network, paths along the river respecting or clashing with the surrounding landscape; natural continuity and nature-urban interactions; presence of any fragmentation; and analysis of stakeholders and their networks. Fifth, the framework includes a study on economic activities, aimed to reconstruct agricultural and rural landscapes, farm profiles and agricultural product inventory, secondary and tertiary activities, and economic imbalance detected in river basins (Fig. 6.3).

This framework, shared by public and private subjects involved in the reclamation process of the river, is the basis from which the river contract starts. It is an integrated analysis concerning environmental, social, and economic aspects of the territory engaged in the river contract process, which contains the starting points to define the operative objectives that different actors must share and actively work to achieve.

This knowledge comes from different sources, such as information officially produced by government bodies, found in the scientific literature, or resulting from oral stories and interviews. Putting all together means that the chemical and physical profile of the river (the quality of the water and the dynamic of its flow, primarily the

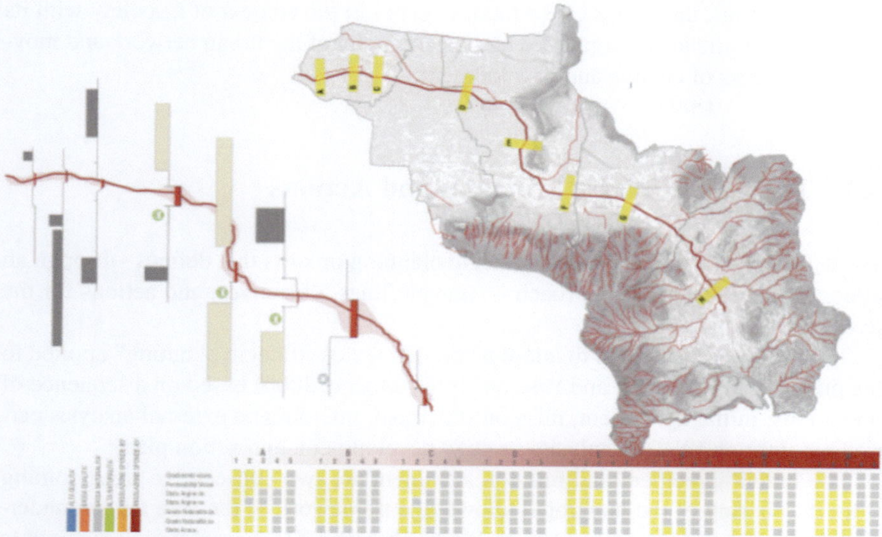

Fig. 6.2 Riverbanks along the Cavaiola stream (Source: Laboratorio Piano Territoriale e Paesaggistico, students: Calderazzo C., D'Alterio S., Di Bonito G., Murolo S.)

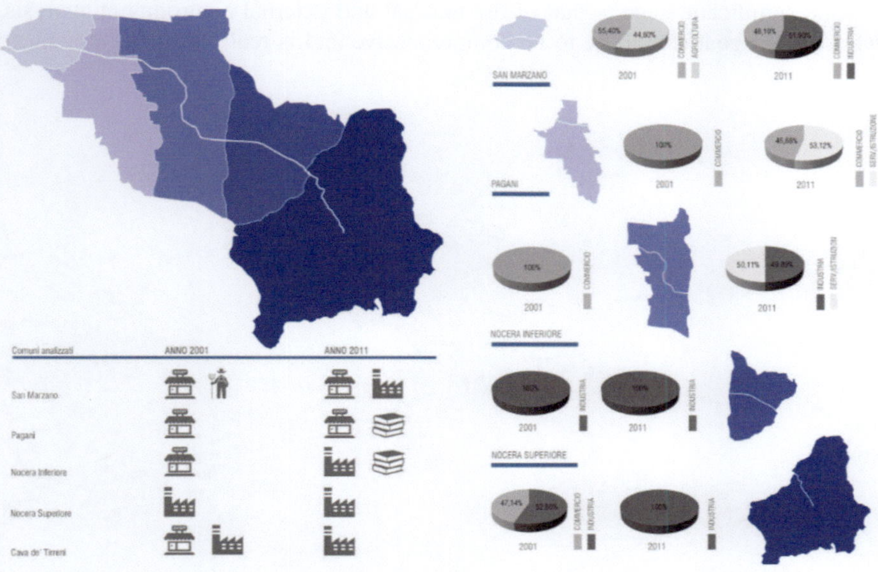

Fig. 6.3 Economic activities in the Cavaiola stream territory (Source: Laboratorio Piano Territoriale e Paesaggistico, students: Calderazzo C., D'Alterio S., Di Bonito G., Murolo S.)

risk of flooding), the center of the picture, is put in the context of a society with its needs and significances, organized in the geography of the urban network and moving in a process of change and development.

6.4 The Action Plan: Strategies and Actions

The action plan is the result of a strategic planning process that defines –through an integrated and concerted approach – strategic lines, objectives, and actions for the river basin's future.

The proposed method is to adopt a "private sector strategic planning" applied to the public sector (Bryson and Roering 1987; Moccia 2000) based on a sequence of operations: initial agreement, mission statement, internal and external analysis carried out through SWOT analysis, strategy development, and action plan.

The initial agreement and mission statement are two consecutive steps coming from the sharing of the integrated knowledge framework of the area and an understanding of relevant resources and problems observed. On these issues, it is possible to start an agreement with a common path in mind; that's the reason why also the methodology to adopt in the process is discussed and shared in this preliminary stage (Fig. 6.4). All the involved actors must be convinced of adopting the chosen approach.

One significant stage is that of the internal and external environment analysis, whose main goal should be to be comprehensive and is really difficult to achieve

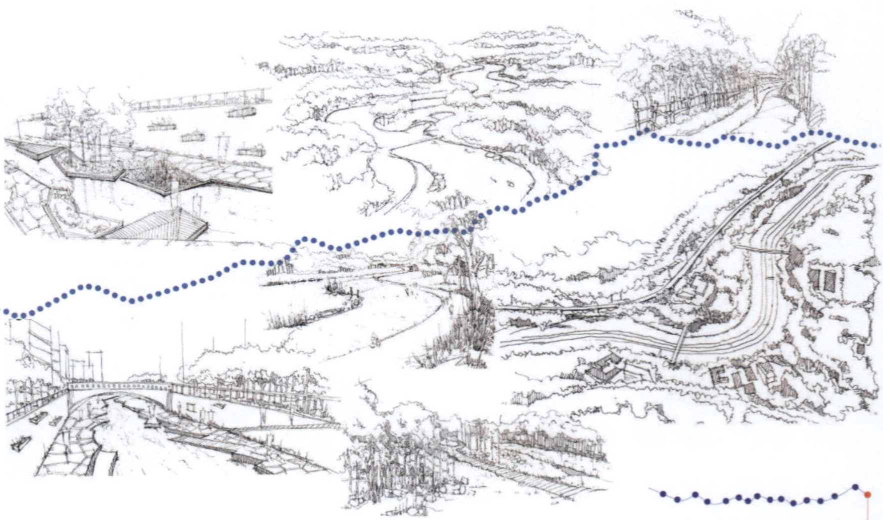

Fig. 6.4 Future scenario for the Irno valley (Source: Laboratorio Piano Territoriale e Paesaggistico, students: Sacco A., HAO W.)

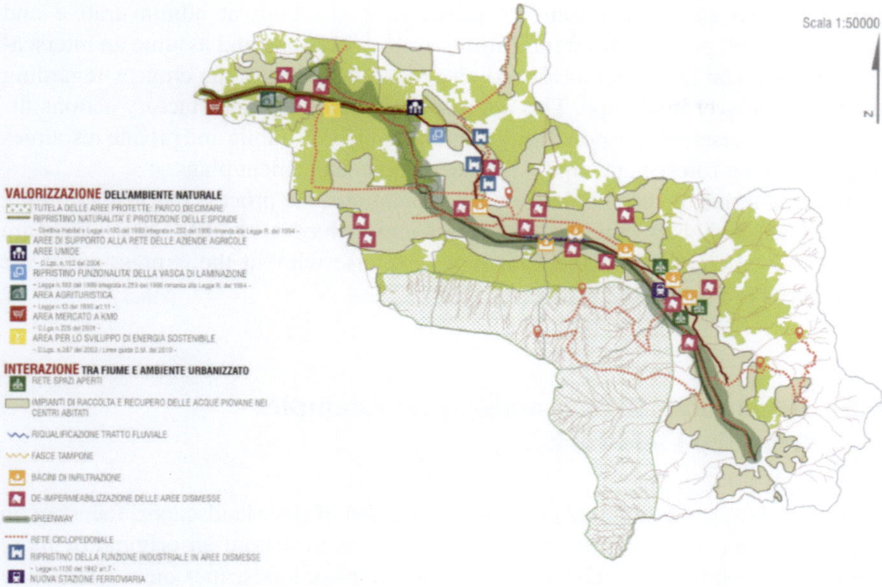

Fig. 6.5 The map of strategies (Source: Laboratorio Piano Territoriale e Paesaggistico, students: Calderazzo C., D'Alterio S., Di Bonito G., Murolo S.)

(Moccia 2005, 21). In the internal environment, strengths and weaknesses of the studied area are pointed out; in the external environment, opportunities and threats concerning the possibility of change of the area are indicated. The unit of measurement for this analysis is given by the mission statement.

The strategic territorial scenario is the support of the strategy development and consists in a large-scale map in which it is possible to read the future development of the area according to the strategies shared by participants in the river contract process (Fig. 6.5). The map aims to indicate the areas subject to urban and landscape change making comprehensible and easy to read spatial and functional relationships among the various proposed strategies in the chosen large-scale section.

Usually action plans focus on the following issues: river landscape fruition, pollution and ecosystem functioning, hydraulic and hydrogeological risk, accessibility, and energy production by renewable sources. Actions pertain not only to the regional scale but concern also detail issues, such as the ordinary maintenance and management of river territories or agreements on land uses.

Also the involved stakeholders, their relative duties and commitments, the time line and the implementation process, and necessary financial and human resources are included in the action plan that participants in the river contract share and commit to carry out signing an agreement.

Adopting an integrated approach, involving citizens and local forces coming from different fields in the river contract process is useful to have a good knowledge of critical situations in the area and prepare effective solutions, but it also increases the awareness of water resource's value. It works both at administrative and territorial level, pushing different institutions to coordinate and assume an interscalar approach and involving interested citizens to take part in the choices regarding the future of river landscape. This approach provides more satisfactory actions fitting local interests and the possibility of mobilizing both public and private resources and combining foreseen changes with adequate management plans.

In the absence of stakeholders, in our simulations, the process has just a training significance and shows the limits that factors involved in the SWOT analysis are only those emerging from the knowledge exercise without the expression of the willingness of responsible actors.

6.5 Reclaiming River Landscapes: Examples from the Fieldwork

River contracts are useful tools for the reclamation of river landscapes, focusing on the twofold meaning of the verb "to reclaim,"[6] as to restore an original or more appropriate condition of ecosystems and to take back landscapes on a community base.

In this framework, reclaiming river landscapes entails to use in the field the theories of landscape restoration and the principles of community-based development[7]. With regard to river landscapes, on the one hand, recurring to landscape restoration might suggest experimentations useful to recover degraded river landscapes pushing them to a contemporary natural condition (Moccia 2013, 2014); on the other hand, focusing on local resources, putting at work together contextual and expert knowledge, might play a significant role in the interpretation and revaluation of river landscapes (Magnaghi 2010; Giacomozzi and Magnaghi 2009).

The proposed approach to landscape restoration assumes the theme of recognizing cities as part of the natural world and designing them accordingly, emphasizing ecological processes: "This is not just a matter of imitating or echoing the shape or natural features [...] but of adapting urban form to natural processes" (Spirn 2012, 8). As proposed by the "pioneering environmental thinker" George Perkins Marsh: "in reclaiming and reoccupying lands laid waste by human improvidence or malice...the task...is to become a co-worker with nature in the reconstruction of the damaged fabric" (March 1865 in Spirn 2012, 2). This challenge is even more difficult to face in the complexity of contemporary cities where it is necessary to respond to the objective of restoring natural cycles in urbanized areas.

Taking back landscapes on a community base means adopting some experiences and tools in the field of the social perception of landscapes (Clifford 1993; Clifford

[6] The verb "to reclaim" comes from the Latin "reclamare" that means "to recover" and "to claim back."

[7] On the effects of an action-oriented approach to research in order to orient community-based natural resource management, see Saija 2014.

et al. 2006) and designing and carrying out surveys (Lynch 1960, 1984, 1990), whose outcomes are able to construct a framework of the relevant issues for the revitalization of river landscapes. This community-based reconstruction supports actions of development focused on the promotion of the sense of places and assures that all the relevant actors are involved in the process. As a matter of fact, when working to set up a river contract, one of the warnings to bear in mind is to pay attention to the role of absents (Bobbio and Saroglia 2008) that, since not involved in the process, do not have responsibilities and bonds to respect and can condition the river contract process.

The main problems we meet in river surveys are linked to the quality of water – the level of pollution and water functionality – risks coming from landslide or flooding, and the need the community has to appropriate landscape and ecosystem services. In river contract action plans, it is quite impossible to find separated responses to problems because projects – and this is one of their quality – mix together solutions with multiple significances and functions.

We tried to synthetize the studied examples, according to some significant issues that we could recognize in the aforementioned simulations of different stages of river contract processes: how to recover river landscapes, floods and dynamic landscapes, riverfront fruition, and river landscape images.[8]

6.5.1 How to Recover River Landscapes

The first issue deals with recovering river landscapes that means to regain the natural functionality of flowing waters and then to harmonize the hydrologic network to the urban system. Key strategies are remediation, renaturalization, restoration, reconversion, reuse, and assuring connections through a network. It is easy to note that there is a clear predominance of a rehabilitation approach aiming to interpret almost hidden traces and clues of the past to explore potential scenarios responding to contemporary challenges.

This approach includes renaturalization projects and the restoration of existing wetlands and the reuse of ancient drainage channels – sometimes abandoned or partially covered by the road network – as greenways connecting different landscapes[9] (Fig. 6.6) and tanks to collect surface waters or, bringing them to surface, as new attractive places for the towns along the river (Fig. 6.7). Also the ecological restoration of disused quarries is a problem to face, together with the design of

[8] It is evident that issues are intertwining and not separate. The choice to distinguish them according to the predominance of one issue on the others is a rhetorical device.

[9] In this framework, an interesting example is given by some sections of the Sarno River in the Campania region of Italy, especially in the area bordering its delta and along the coast where the Conte di Sarno drainage channel is located.

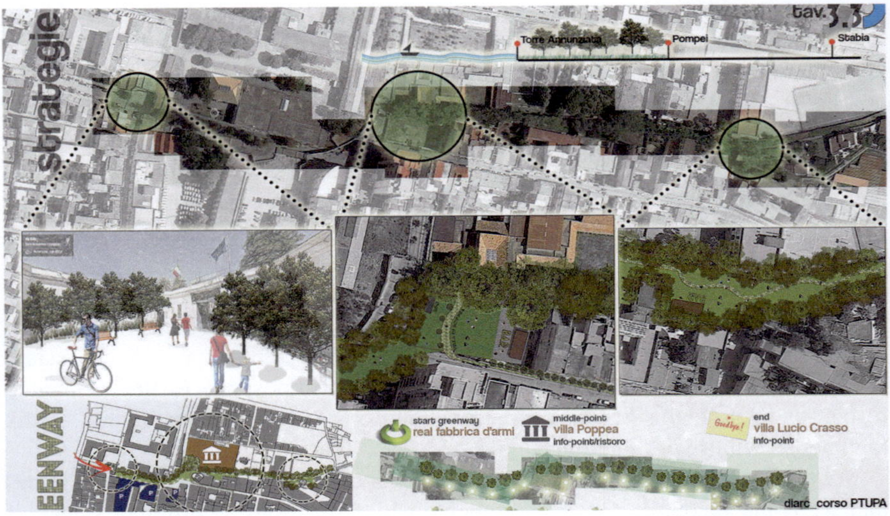

Fig. 6.6 The greenway as a strategy to reuse an ancient drainage channel)

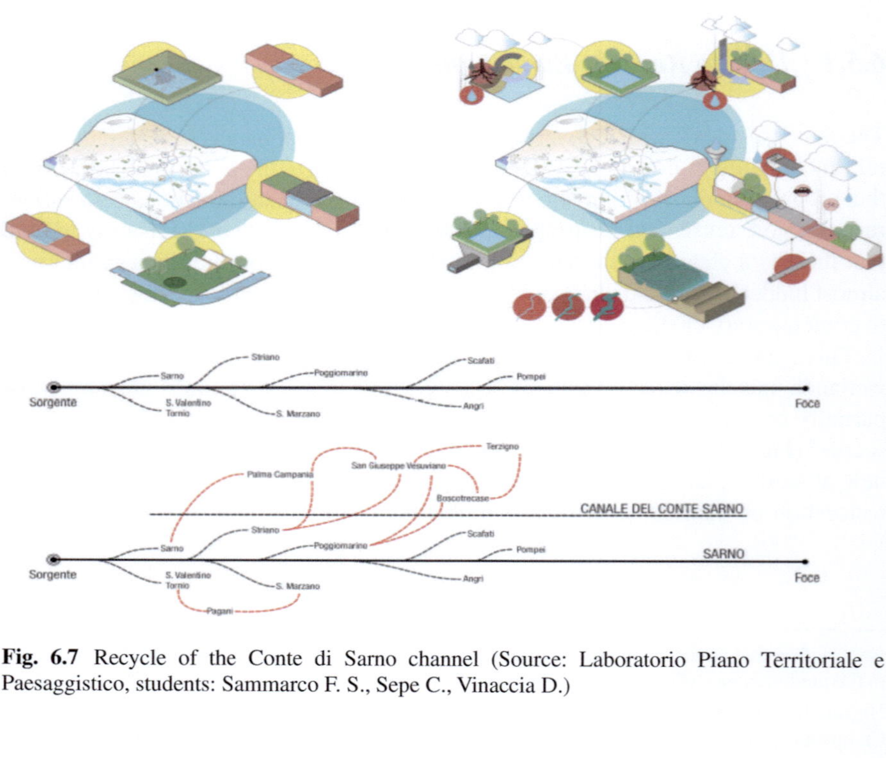

Fig. 6.7 Recycle of the Conte di Sarno channel (Source: Laboratorio Piano Territoriale e Paesaggistico, students: Sammarco F. S., Sepe C., Vinaccia D.)

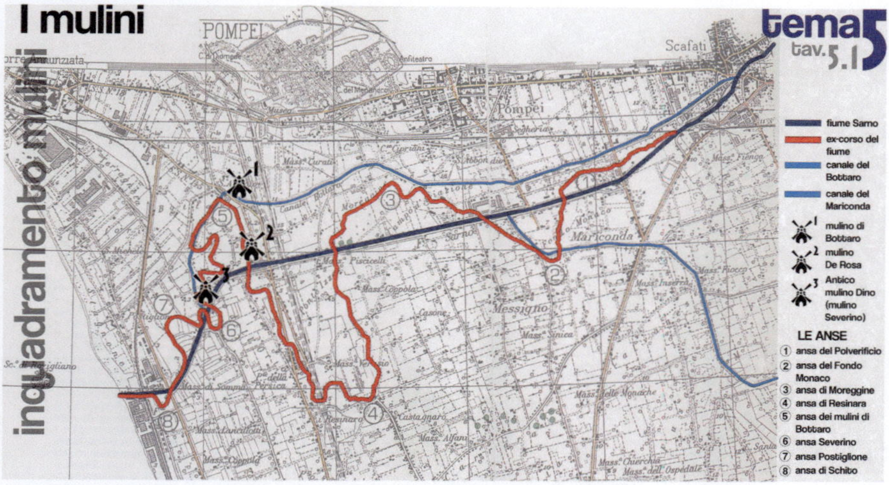

Fig. 6.8 The network of mills in the Sarno plain (Source: Laboratorio Piano Territoriale e Paesaggistico, students: Esposito F., Maldifassi G., Migliardi L., Oliva G., Palermo G.)

protected areas, educational paths, and the connection to greenways and the cycle route network.[10]

There are often abandoned mills along many river corridors that can become a resource not only to recall a thriving past in which the river was a power source for bordering territories but also to be reconverted to a new use. The network of mills can be read as a system of landmarks in a fluvial park, reused as alternative energy resource[11] (microhydro), in which the river works as landscape connector (Fig. 6.8).

Sometimes, entire neighborhoods have been abandoned over the years, due to natural causes or to protect from risks, causing the town to develop elsewhere. If the neighborhood was characterized by the presence of tanneries or other potentially pollutant industries taking advantage of the river,[12] a restoration project could be articulated according to the following lines of action: remediation, to remove sources of pollution and pollutants in the soil and in surface and underground water, riverbed cleaning and banks nourishment, and culture, in order to promote the town's rebirth. As for the tanneries, a shift to a natural tanning system could be advanced, with an attention to the reuse of the dross of the process.

[10] It is the case of the river contract of a section of the basin of the Po River in Emilia-Romagna that focuses on the ecological restoration of the quarry known as "Lanca dei Francesi," in the town of Roccabianca (Parma).

[11] This strategy is adopted in the territory around Scafati in the Sarno plain and by the northern Tordino valley river contract.

[12] A simulation of river contract focuses on the restoration of the Toppolo district, in the territory of Solofra, where a tributary of the Sarno River called Solofrana flows, completely abandoned after the earthquake of 1980, when the tanneries that constituted the heart of the economic and social life of the town were delocalized in the industrial area downstream.

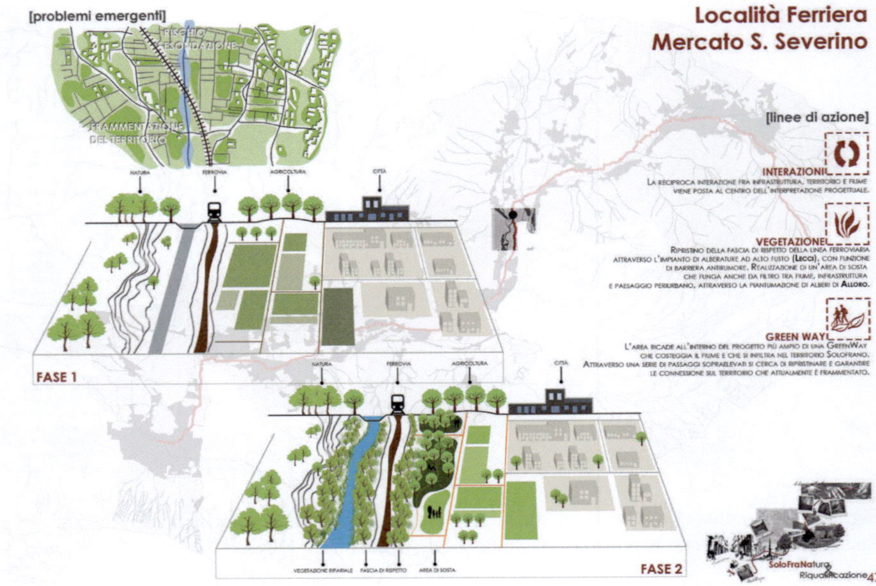

Fig. 6.9 The railway line and the Solofrana stream as landscape connectors (Source: Laboratorio Piano Territoriale e Paesaggistico, students: Avallone A., Cinque M. R., Gabrielli M., Senatore A.)

Recovering river landscapes often entails adopting a sustainable mobility and green infrastructure approach, directed to restore and assure connections in a fragmented territory (Fig. 6.9).

Contemporarily, such an approach helps to enhance existing and potential resources linking them in a network: archeological excavations, contemporary fragments of landscape characterized by a mix of incomplete works and domesticated nature advances, clues of pilgrimages along riparian corridors, and memories of an ancient fluvial port completely erased by the expansion of the area.[13] Often, there is no connection among different resources, and some of them are almost unknown. River contract simulations plan to construct ecological and landscape networks sustained by tourism strategies able to respond to a wide range of demands and to develop spare time activities compatible with agrarian landscape, transforming the area in an archeological agrarian fluvial park. Sometimes, however, the main strategy consists in connecting river territories, through an overlapping layer system: the

---

[13] For instance, from an archeological point of view, the Sarno river basin is very rich, both in the area of the springs and of the delta. It is the case of Longola prehistoric settlement (Albore Livadie et al. 2010), in the town of Poggiomarino where it is possible to experiment a not taken for granted vision between archeology and monument to the incomplete, as the works to realize a purification plant started in the area and then were blocked for the presence of archeological excavations. Also Pompeii – destination of composite pilgrimages directed to the ancient Roman city excavations (Maiuri 1975) and to the shrine of the Virgin of the Rosary – is not far from the Sarno River and constituted its fluvial port.

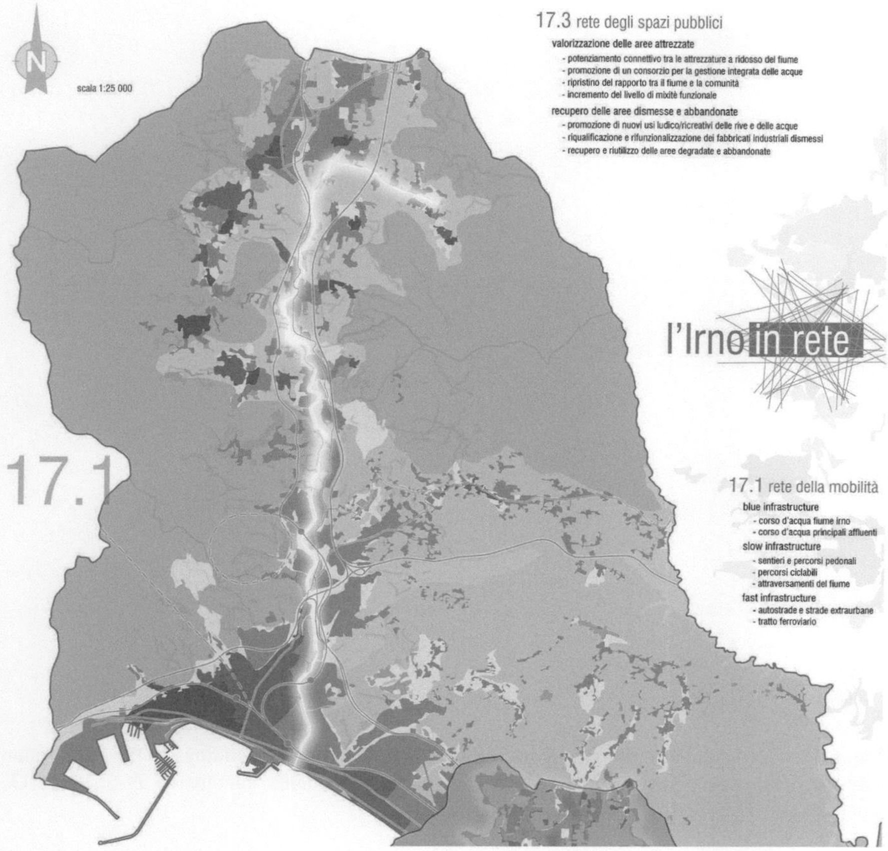

Fig. 6.10 The three networks along the Irno River: the mobility network, the network of public spaces, and the agricultural production network (Source: Laboratorio Piano Territoriale e Paesaggistico, students: Conte L., De Chiara A.)

mobility network, articulated in blue, slow and fast infrastructure; the network of public spaces, including new equipped play areas and the recycle of abandoned and disused areas; and the agricultural production network, with the aim to promote local products, supporting the institution of a consortium and a brand, mainly focusing on adaptive farming (Fig. 6.10).

6.5.2 Dynamic Landscapes

The second issue concerns responses to problems caused by frequent floods pervading riverfront territories. These events entail the necessity to design dynamic landscapes with variable features over time according to current weather and climate

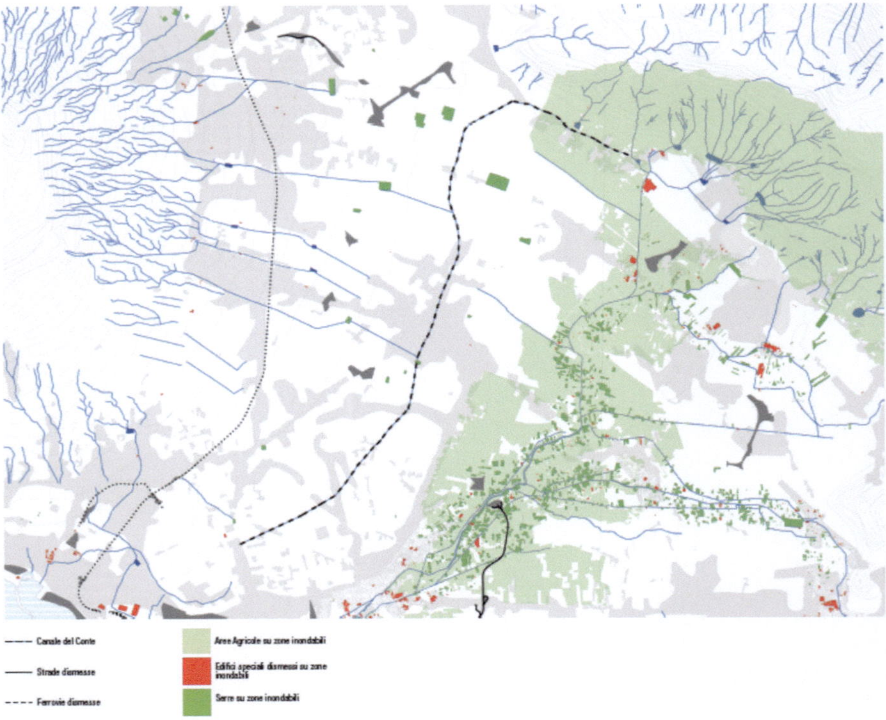

Fig. 6.11 Flooding areas, the abandoned railway line, and disused buildings in the Sarno plain (Source: Laboratorio Piano Territoriale e Paesaggistico, students: Sammarco F. S., Sepe C., Vinaccia D.)

conditions, starting from the porosity of river corridors. Dynamic landscapes put at stake in a site-specific way the relationship between resilience and resistance and pose the question of how to redesign the borders between spaces able to adapt to the presence of water and spaces destined to remain unchanged in the course of time.[14]

Stormwater management and recycling and the increase of soil permeability are crucial not only from the point of view of a return to nature (even if urban) but also as a natural and not artificial response to flood risks, counting on ecosystem services. A new paradigm has to be applied, overcoming the "end-of-pipe-control" paradigm that adopts fast conveyance systems: "the precipitation – runoff –

[14] On the relationship between resilience and resistance, the reference is Bernardo Secchi's groundbreaking essay "The conditions are no longer the same," in which he invites us to select simple relationships: "to separate what is, in the towns and in the territory, hard from what is soft and malleable in its properties, its physical asset, its functions, and in its relation with other objects and in its overall sense. […] Hard and soft are terms which describe not only physical properties and visual relations. […] Hard and malleable are terms close to negotiable not negotiable" (Secchi 1984, 12). On the potentials of the porous city, see also Secchi (2007), Secchi and Viganò (2012).

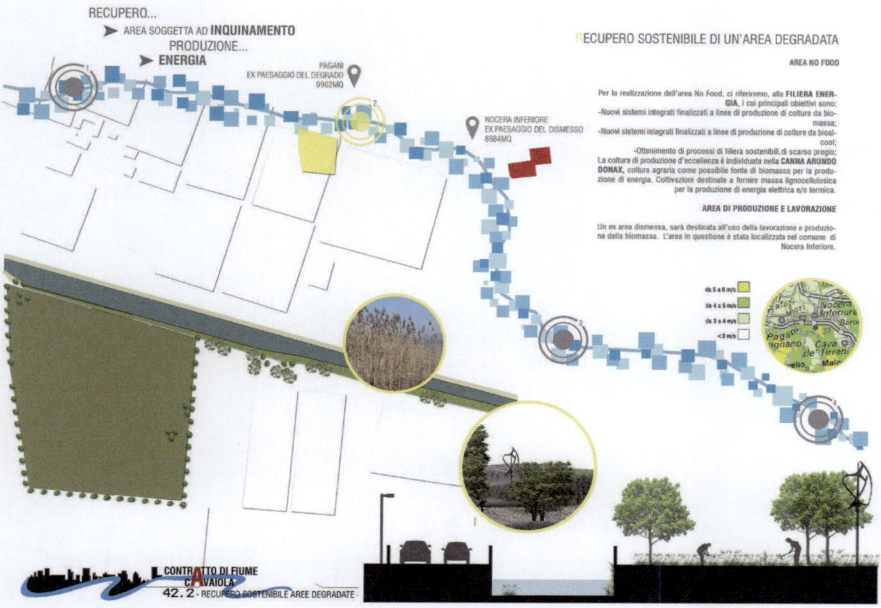

Fig. 6.12 No food cultivations along the Cavaiola stream

groundwater recharge balance of the cities of the future can approach the natural hydrologic cycle" (Novotny 2008, 7).

Different approaches can be adopted depending on the presence of agrarian or urbanized landscape where floods occur.[15] Actions include, for agricultural areas affected by strong hydraulic risk, the development of a succession of controlled flooding green areas. However, as for the presence of flooding in urban settings, solutions focus on the redesign and renaturalization of river banks together with a stormwater management approach taking measures to increase soil permeability (Fig. 6.11). In the presence of high-risk zones, due to the bad state of banks and riverbed, strategies involve to restore the naturality of banks through low-impact measures (for instance, naturalistic engineering works).

When lands cannot be cultivated because of their recent use harmful to the environment (industries, warehouses), a possible scenario envisages the construction of

[15] A striking example in river contract simulations regards, in the Sarno river basin, the critical situation of via Ripuaria in Pompeii. Rainfalls cause heavy floods that make the road network useless, causing to Torre Annunziata and Castellammare di Stabia's inhabitants to remain isolated. The strategy, founded on a sustainable water management approach, consists in delocalizing the road network adopting an alternative solution; modifying, widening, and naturalizing river banks; and transforming via Ripuaria in a greenway.

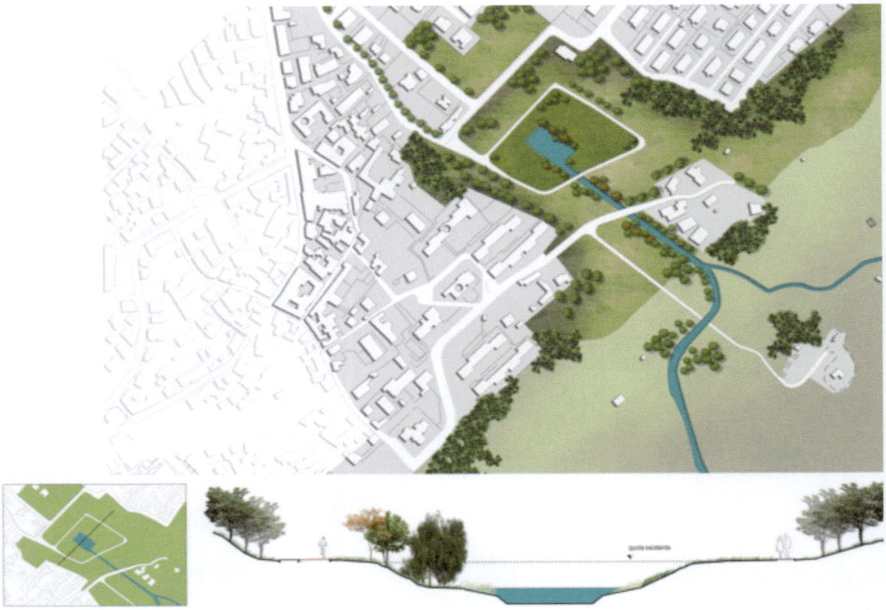

Fig. 6.13 Strategies to make room for the river in eastern Naples (Source: Master thesis, Claudia Vegezio)

artificial wetlands[16] for absorbing chemical pollution (Fig. 6.12), thus combining both the beauty of form and the ecological function. In riverfront polluted areas, it is possible to plant no food cultivations, using soils to produce energy. Moreover, in order to restore the relationship between the river and riverfront territories, infiltration basins can be promoted.

The strategy of giving back space to the river is adopted in case of highly urbanized urban regions that were originally marshlands, characterized by many hydraulic risk areas.[17] Useful actions for an ecological water-centered design approach (Berruti et al. 2013) are connecting channels and renaturalizing buried ones when possible, reinforcing existing paths and designing new ones, and working to bring people in contact with water (Fig. 6.13). This approach pursues the well-being of citizens and the vitality of the territory, not only avoiding the occurrence of risks but also actively cooperating with the surrounding environment.

[16] Wetlands work as sponges, collecting water during floods and giving it back in periods of low water; they constitute also tanks for aquifers and ideal habitat to maintain biodiversity. For a guide to wetland design, cf. France 2003.

[17] In the example of the partially buried Molaro channel in the Sebeto plain, at the boundaries between the towns of Massa di Somma and Pollena Trocchia, in eastern Napoli, the proposal involves bringing the part of the channel that lies underneath via Veseri back to surface. Open spaces adjacent to the channel have been destined to receive part of the water as an infiltration basin. Then, new routes suitable for vehicle and bikes are designed.

6.5.3 Riverfront Fruition

The third issue deals with the natural functioning of river corridors, their adaptation to natural processes, and the attention to vitality in its various aspects, from the good biologic functionality to the capacity of activating livable landscapes.

In contrast with the artificiality connected with hard engineering methods treating river landscapes' problems, river contract simulations focus on developments based on ecological infrastructure and on a mixed and functional dimension of landscape, which define healthy and vital places. In this perspective, rivers are not considered marginal anymore: they are not used for wastewater disposal or polluted with the dross of agricultural and productive processes.

Riverfront areas can help to respond to new demands for enjoyment of landscape, able to allow knowing nature and its processes, open-air sport activities, and ecotourism, contributing to the local development of adjacent urban regions. River parks, bike trails, outdoor games facilities, and sport facilities are examples of possible realization in the framework of riverfront fruition, also catching the opportunities connected with the reuse of vacant buildings and lots in surrounding areas ("blue voids," Desimini 2013).

Strategies entail promoting the scenarios of living together, from the river festival to the trout fair, pointing on activities such as rafting and fishing, and working for the reuse of former abandoned buildings (Fig. 6.14). As for sustainable mobility actions, the river has to become the backbone of collective activities; it is necessary to take river banks back to the public and also to recall old local traditions (Fig. 6.15). To establish a green and blue corridor can help connect river mosaics through an ecological, environmental, and functional approach and can overcome fragmentation[18] (Fig. 6.16).

Along disused railway lines, an eco-urban scenario can rely on the structural role of an innovative system of public transport.[19] Food and agriculture parks can be promoted, with a productive and educational intent, hosting several activities from cultivation and transformation to production and sale, and the organization of workshops focused on the different stages of the food production chain.

---

[18] As far as the Tanagro River is concerned, which flows for a long section in mountain area, the main problem to face is accessibility: first of all mountains, then the Napoli-Reggio Calabria highway works as barriers to reach the river. Moreover, the area lacks a local road network close to the river, and no riverfront facilities are provided. In the case of the Fibreno river contract in the Liri Garigliano basin, the aim is to reclaim riverfront territories through a green corridor respecting natural resources and enhancing a responsible tourism. In the example of the Cavaiola, the objective is to trigger "slow intensities" underlining the "pool and riffle" structure of river corridors and exploiting their porosity.

[19] In the Alento spring area, historical centers are destined to become the door to the other towns of the plain. Improving tourism and economic development of adjacent territories is the main strategy of the contract of the Biferno River, in the Molise region, where actions to improve slow mobility combine with the organization of several sport activities, especially in the area of the Guardialfiera Lake.

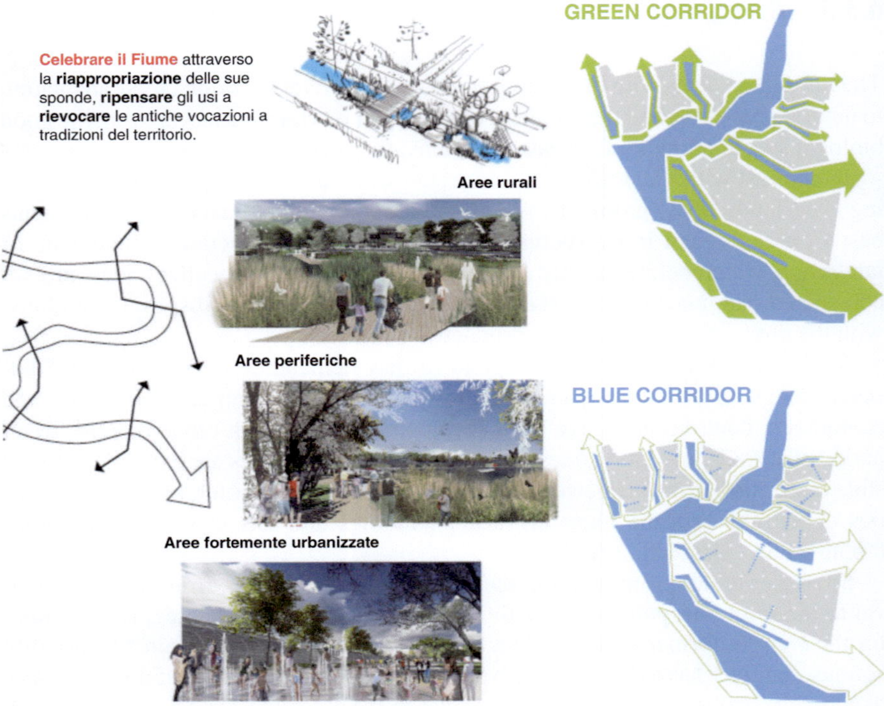

Fig. 6.14 The scenarios of living together in the Sarno plain (Source: Laboratorio Piano Territoriale e Paesaggistico, students: Sammarco F. S., Sepe C., Vinaccia D.)

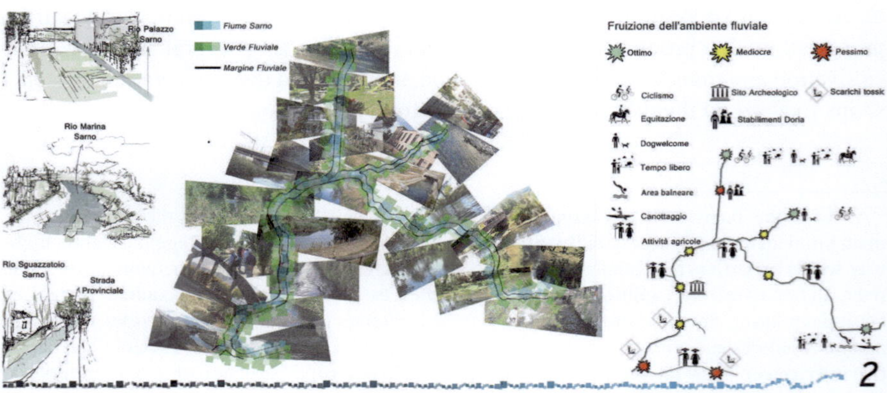

Fig. 6.15 River bank uses and Fruition Assessment in the Sarno River basin (Source: Laboratorio Piano Territoriale e Paesaggistico, students: Borriello V., Molaro C., Saviello A.)

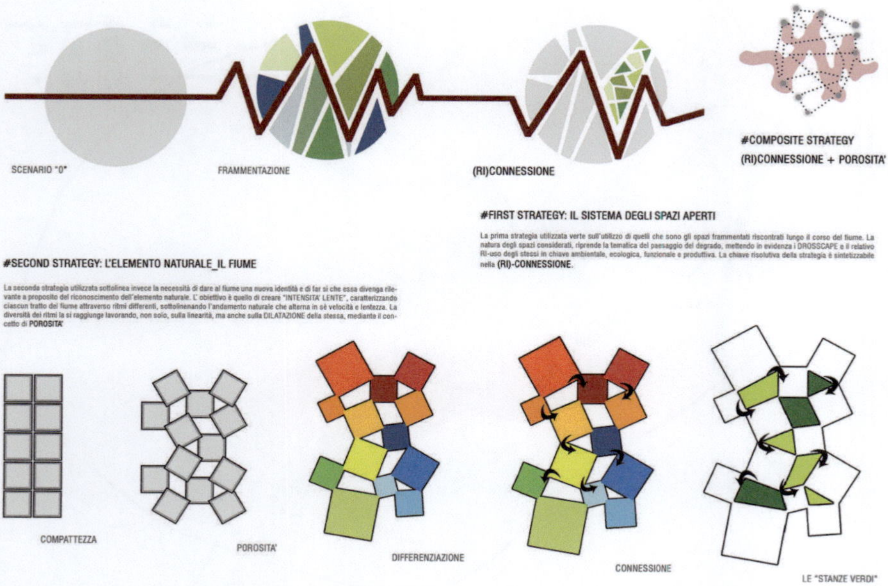

Fig. 6.16 A composite strategy for the Cavaiola stream: reconnecting porous places (Source: Laboratorio Piano Territoriale e Paesaggistico, students: Calderazzo C., D'Alterio S., Di Bonito G., Murolo S.)

6.5.4 River Landscape Images

Relevant issues for the revitalization of river landscapes can be also deduced from a survey focused on river landscape images, inspired by Kevin Lynch's "image survey" (1960, 1984, 1990) and adapted to the exploration of an urban region. The survey aims to analyze how river landscapes are perceived, adopting the "five elements"[20] in which people structure their perception of cities.

In this framework, the functioning of the river is investigated together with the existing relationships between inhabitants and the river. The investigation also puts ecosystem services (or, more often, "disservices," Sagie et al. 2013) to the test and studies local people's availability to activate in order to support river landscape values. Recurring components of river image, existing bonds with the river, comparison with the situation of the past, problems undermining the relationship with the river (pollution, insufficient accessibility, lack of care and maintenance, scarcity of facilities for riverfront fruition), and routines undertaken by inhabitants are some

[20] The survey aims to identify which is the public image, or maybe which are the public images, of an urban region. We are dealing with a different scale image compared to the classical image of the city. For a survey on community resilience also inspired by Lynch's theories, see Palestino 2013.

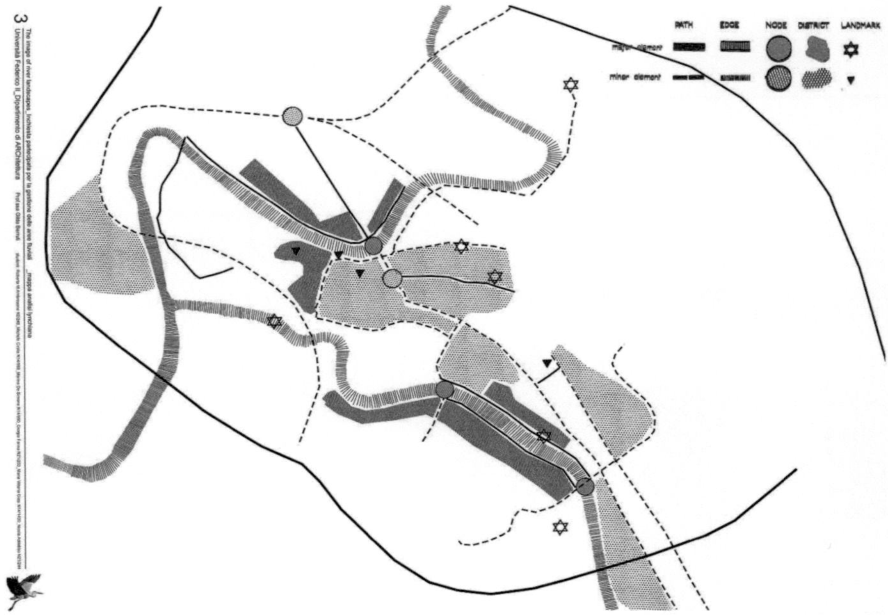

Fig. 6.17 The image of Calore and Sabato river landscapes in Benevento (Source: Corso "The image of river landscapes. Survey on the management of riverfront territories" students: Ambrosone R., Cirella M., De Simone M., Farina G., Giasi M. V., Addabbo N.)

of the elements emerging from the survey[21] (Fig. 6.17) that could be at the heart of a landscape restoration process (Moccia 2013, 2014) of river territories. As a matter of fact, the survey points out clues of the past and fragments of historical memories and configurations to bear in mind to go toward a future respectful of the sense of places, in which co-working with nature entails mixing together different elements, in a composite process.[22]

The final outcome of the survey reports relevant issues to consider in order to direct the river revitalization process and an outline of actors to involve in the process. As for the relevant issues, the survey results also make possible to become

[21] An interesting survey concerns the Calore Irpino River and its tributary Sabato in the section corresponding to the town of Benevento. The perceived images of the two rivers are fragmented. Rivers are perceived mainly as places of transit, to be crossed to reach the prearranged destination. Bridges are also the borders of the lived and described districts that can be read observing how the town is used. There is an emotional bond with the river and the idea of the river goes with a kind of fear caused by the belief that it can become dangerous.

[22] An example is given by the simulation on the Solofrana stream, in the Sarno river basin, that today is reduced to an artificial stream nourished by the drains coming from the surrounding industrial areas and towns. It is a "ghost river," in part buried, that lost the character of vital element for the territory, so definitively destroying the relationship with the water and also compromising the control of the stream. The restoration of the Toppolo district in Solofra involves a return to a natural tanning system, able to reuse the dross of the process (biogas, biomasses, and organic fertilizers).

aware of which is negotiable and which is not in the process of revitalization of river landscapes, reasoning on which are "hard and malleable" places and issues.

The outline of actors allows designing a framework of positions regarding riverfront territories, to be able to understand balances and potential conflicts and control risks related to absent actors' agency, strongly affecting the process just because they decided not to take part. This means involvement of those who have been both accidentally excluded and handling free agents, with no commitments, but are able to condition the success of the process.

Finally, this community-based survey might have the added value of recognizing the value of "narrating resilience" (Goldstein et al. 2015), adopting a storytelling approach which makes possible not expected narratives about places, working as hints to change the current situation[23].

One of the main fields of application of this approach is to bring the river back to be part of the city, as the symptomatic case of the Los Angeles River clearly shows (Gottlieb and Azuma 2007). This experience can be a lesson to many rivers being in a terrible state[24] that, nevertheless, show a reasonable presence of groups and organizations reclaiming them and a collection of stories, locally stratified, to release and to narrate.

6.6 Conclusions

River contracts are becoming a planning practice speedily spreading in Italian regions. Their background is in many other programs where a participatory approach obtained the involvement of actors in a collective effort to reach a common agreement on objectives. These experiments were first developed in the economic field with territorial pacts (Scassellati 1998; Moccia 2002), had subsequently other regional and territorial impacts, and became a way to innovate regulatory and master plans.

The idea of pacts, in other words, contracts that join together public and private organizations, is always more frequently adopted by public administrations in Italy, taking the place of hierarchical top-down decision processes and implementation directives. In addition to a large democracy in decision-making, it distributes responsibilities to a larger set of actors, making citizens part of the governance. This mobilization of human capital and all the other resources connected to it (expertise, economic, organization) is a chance that makes plans more effective, this being one

[23] It is the case of "storytelling as a model *for* planning" (van Hulst 2012), in which stories are used as "tools," "in the service of change, as shapers of a new imagination of alternatives" (Sandercock 2003, 9).

[24] The Sarno River could be an interesting case in which to experiment a "narrating resilience" approach, starting from local associations and groups that are small portions of a latent network that can be linked in a shared path toward a healthy development of the river. For a close examination of the results of the survey on the images of the Sarno River as hints for action, see Berruti (2016) and on resilient river landscapes Berruti (2014).

of the most felt requirements called to overcome the weak results of the past planning exercises.

The second expectation that fuels the spread of river contracts in Italy is the necessity to resolve the fragmentation of responsibilities of public bodies on complex environmental problems as well as over many other issues concerning local government. In many cases, decision-making is an interinstitutional process in a public administration used to separate operations and, frequently, jealous of its autonomy, at worst, willingly to defend its decisional power from exterior interference. Lacking an authority to make a synthesis or enforce decisions from the top in the name of a superior public good, because a plurality of interests is now legitimized, cooperation is needed among the many representatives of such interests.

The more striking innovation of river contracts is their appearance in the ecological field where more absolute ideologies seemed dominating. Instead of a comprehensive and regulatory planning, they propose strategic actions, not in contraposition neither with the intent of replacing the system of protection of natural resources but adding to those achievements more effective programs.

As a matter of fact, the degradation of the natural environment and the ecosystems has continued, but the climatic challenge has become so hard that working toward the preservation of existing natural resources is not enough. That is why, also in the planning field, experiments of innovation are important, and the now starting movement for environmental contracts is a practice that deserves scientific attention and accurate research during its making.

In conclusion, the role of research, as explored in our planning studios and investigation, is to support participants to the discussions and to the planning process in general with a basic knowledge; to pick up and collect pertinent information from multiple sciences; to translate them in a common language and combine the same with people knowledge and oral stories; to design a tailor-made planning process that fits the place, the stakeholders, and the time; to show, on the basis of new paradigms and best practices, potential changes and their benefits; and to articulate desired outcomes in strategies and projects, so that they can profit from available resources and public financing, having in mind the necessity of monitoring and control.

References

Albore Livadie C, Castaldo E, Castaldo N, Cesarano B, Citro D, d'Avella A, delle Donne M, Pappalardo M T, Pizzano N, Vannata R (2010) Le strutture abitative e di servizio dell'insediamento dell'età del ferro di Longola (Poggiomarino, Na), Atti IX Incontro di Studi Preistoria e Protostoria in Etruria, 12–14 settembre 2008, Pitigliano/Valentano. Centro Studi di Preistoria e Archeologia, Milano, pp 539–551

Berruti G (2014) Paesaggi fluviali resilienti Re-imagining the Los Angeles River. In: Palestino MF (ed) Spazi spugna. Esperienze di pianificazione e progetto sensibili alle acque. CLEAN, Napoli, pp 67–84

Berruti G (2016) Bringing river landscapes back to life. Notes from a survey on the Sarno river, UNISCAPE En-Route 3:119–126, (online). http://www.uniscape.eu/allegati/UNISCAPE%20En%20Route%20n.%203.pdf. (July 2016)

Berruti G, Coppola E, Moccia FD (2013) Urban morphology and ecological water-centered design in Mediterranean areas. In: Moccia FD, Palestino MF (eds) Planning stormwater resilient urban open spaces. CLEAN, Napoli, pp 162–177

Bobbio L, Saroglia P (2008) Lungo il fiume delle politiche contrattualizzate. Dall'esperienza francese al caso piemontese, (online). http://www.regione.piemonte.it/oss_riforma/dwd/semplificazione/lungo_fiume.pdf. (December 2015)

Bryson J, Roering WD (1987) Applying private-sector strategic planning in the public sector. J Am Plan Assoc 53(1):9–22

Clifford S (1993) Places: the city and the invisibile. Public Art Development Trust, London

Clifford S, Maggi M, Murtas D (2006) StrumentIRES. Perché, quando e come realizzare una mappa di comunità. Istituto di ricerche economiche e sociali del Piemonte, Torino

Davoudi S (2012) Resilience: a bridging concept or a dead end? Planning Theory and Practice 13(2):299–307

Desimini J (2013) Blue voids: stormwater strategies for abandoned lands. J Landsc Archit 8(2):64–73

EU Directive 2000/60/EC (2000) Framework for Community action in the field of water policy. Official Journal (OJ L 327), Bruxelles

Forman RTT (1995) Land mosaics the ecology of landscapes and regions. Cambridge University Press, Cambridge, MA

France RL (2003) Wetland design. Principles and practices for landscape architects and land-use planners. WW Norton, New York

Gandy M (2004) Rethinking urban metabolism: water, space and the modern city. City 8(3):363–379

Giacomozzi S, Magnaghi A (2009) Un fiume per il territorio: indirizzi progettuali per il Parco Fluviale del Valdarno Empolese. Firenze University Press, Firenze

Goldstein BE, Taufen Wessels A, Lejano R, Butler W (2015) Narrating resilience: transforming urban systems through collaborative storytelling. Urban Stud 52(7):1285–1303

Gottlieb R, Azuma AM (2007) Bankside Los Angeles. In: Kibel PS (ed) Rivertown rethinking urban rivers. The MIT Press, Cambridge, MA, pp 23–46

IPCC (2014) Climate change 2014. Impacts, adaptation and vulnerability. Part A: global and sectoral aspects. Cambridge University Press, Cambridge, MA/New York

Italian Law n. 221/2015 (2015) Disposizioni in materia ambientale per promuovere misure di green economy e per il contenimento dell'uso eccessivo di risorse naturali. Gazzetta Ufficiale n. 13 of January 18th 2016. Istituto Poligrafico e Zecca dello Stato, Rome

Italian Legislative Decree n.152/2006 (2006) Norme in materia ambientale. Gazzetta Ufficiale n. 88 of April 14th 2006. Istituto Poligrafico e Zecca dello Stato, Rome

Lynch K (1960) The image of the city. The MIT Press, Cambridge, MA

Lynch K (1984) Reconsidering the image of the city. In: Banerjee T, Southworth M (eds) City sense and city design. The MIT Press, Cambridge, MA, pp 247–256

Lynch K (1990) A process of community visual survey. In: Banerjee T, Southworth M (eds) City sense and city design. The MIT Press, Cambridge, MA, pp 263–286

Magnaghi A (2010) Il progetto locale. Verso la coscienza di luogo. Bollati Boringhieri, Torino

Maiuri A (1975) Pompei. Istituto Poligrafico dello Stato, Roma

Moccia FD (2000) Comuni sotto pressione. La risposta della pianificazione strategica – City hall under pressure: the strategic planning answer. Urbanistica LII 115:153–159

Moccia FD (2002) Patti territoriali come agenzie permanenti per lo sviluppo locale. Il caso del PTO del nord-est di Napoli. In: Belli A (ed) Il territorio speranza. Politiche territoriali possibili per il Mezzogiorno d'Italia. Alinea, Firenze, pp 197–210

Moccia FD (2005) Pianificazione strategica. In: Moccia FD, Mesolella A (eds) Introduzione alla pianificazione dello sviluppo locale. CLEAN, Napoli

Moccia FD (2013) Ecological restoration methodology. In: Moccia FD, Palestino MF (eds) Planning stormwater resilient urban open spaces. CLEAN, Napoli, pp 12–17

Moccia FD (2014) Restauro paesaggistico ambientale. In: Petroncelli E (a cura di) Progetto paesaggio tra letteratura e scienza. Liguori, Napoli

Novotny V (2008) A new paradigm of sustainable urban drainage and water management". In: Proceedings of the Oxford roundtable workshop on sustainability, Oxford University, August 10–15, pp 1–27

Palestino MF (2013) A survey on community resilience. In: Moccia FD, Palestino MF (eds) Planning stormwater resilient urban open spaces. CLEAN, Napoli, pp 131–145

Sagie H, Morris A, Rofè Y, Orenstein DE, Groner E (2013) Cross-cultural Perceptions of ecosystem services. A social inquiry on both sides of the Israeli-Jordanian border of the Southern Arava Valley Desert. J Arid Environ 97:38–48

Saija L (2014) Proactive conservancy in a contested milieu: from social mobilisation to community-led resource management in the Simeto Valley. J Environ Plan Manag 57(1):27–49

Sandercock L (2003) Out of the closet: the power of story in planning. Planning Theory and Practice 4(1):11–28

Scassellati A (1998) La cultura dello sviluppo locale. Intervista a Ubaldo Scassellati. In: De Rita G, Bonomi A (eds) Manifesto per lo sviluppo locale. Dall'azione di comunità ai Patti territoriali. Bollati Boringhieri, Torino

Secchi B (1984) Le condizioni sono cambiate. The conditions are no longer the same. Casabella 298-299:8–13

Secchi B (2007) Section 1: wasted and reclaimed landscapes: rethinking and redesigning the urban landscape. Places 19(1):6–11

Secchi B, Viganò P (2012) La ville poreuse. Un projet pour le Grand Paris et la métropole de l'après-Kyoto. Metis Presses, Geneve

Spirn AW (2012) Ecological urbanism: a framework for the design of resilient cities, pp 1–35. (online). http://www.annewhistonspirn.com/pdf/Spirn-EcoUrbanism-2012. (December 2015)

Van Hulst M (2012) Storytelling, a model of and a model for planning. Planning Theory 11(3):299–318

Voghera A (2015) River contracts in Italy. An experience for river Management, Recent advances in environmental and earth sciences and economics, (online). http://www.inase.org/library/2015/zakynthos/bypaper/ENG/ENG-58.pdf. (January 2016)

Voghera A, Avidano V (2012) Contratti di fiume. Una proposta metodologica per il torrente Tinella, nel quadro delle esperienze italiane. Archivio Studi Urbani e Regionali 103:42–65

Chapter 7
Design Along the River as a Landscape Matter

Paolo Mellano

Abstract The housing and landscape issues are very closely interconnected. Every architecture project interacts with the landscape, and once it is built, it is an integral part of that landscape.

The task of architecture – and not just that with a capital A – is not simply to make the world more beautiful but above all to help man *live on the earth*, to carve out the best spaces and trajectories for his daily activities – living, studying, working and also having fun, relaxing and spending free time pleasantly.

The town-river dynamic is a circular story and it doesn't always have a happy ending. This dynamic is different in each single town and city. And in Turin, with its four rivers cutting through the city, there are four different scenarios which are very different from one another: the Po, Sangone, Dora Riparia and Stura di Lanzo. Rivers pass through urban districts in very different ways, take on multiple identities and create ever-changing landscapes.

Designing buildings along rivers is a matter of *landscape* and *resilienc*e because it is a subject which requires consideration of the potential dynamics and coherence between the various elements in the built environment, i.e. between the urban fabric and water courses: settlement contexts which have to continually adapt to environmental conditions to acquire the ability to exert multiple actions (productive, economic and working, organisational and maintenance) with the aim of making the area more and more hospitable and welcoming.

Keywords Landscape • Architectural design • Environment • Sustainability • Resilience

P. Mellano (✉)
Department of Architecture and Design, Politecnico di Torino,
Viale Mattioli 39, 10125 Turin, Italy
e-mail: paolo.mellano@polito.it

R. Ingaramo, A. Voghera (eds.), *Topics and Methods for Urban and Landscape Design*, Urban and Landscape Perspectives 19,
DOI 10.1007/978-3-319-51535-9_7

7.1 Introduction: The Landscape Matter

A great deal of theoretical work has now been written on the meaning of *landscape*. Everyone has something to say on the subject it seems. At the same time, however, the landscape (and not just in Italy) is in decline: pollution, deforestation, the ozone hole, climate change, desertification, uncontrolled city expansion and more and more frequent environmental disasters. These are just some of the problems facing cities and which everyone is trying to find a solution for.

The landscape has thus become a critical issue for architects too because housing and landscape issues are very closely interconnected. Every project we build interacts with the landscape, and once it is built, it is an integral part of that landscape.

It is not my intention to reiterate the notions and definitions set out previously and referred to elsewhere here (Isola 1991, 1993a, 1995; Bruna and Mellano 2006, 2011), but it is important to underline once again that it is precisely because we live in and on the landscape that it is crucial that we look after it and put it first and foremost in our priorities.

Every architecture project interacts with the landscape, and once it is built, it is an integral part of that landscape. '[...] architecture is a collection of modifications and alterations to the earth's surface for the purposes of satisfying human needs [...]' (Morris 1881).

Rivers are generally at the heart of a settlement's raison d'être, and the relationship which the latter establishes with its water course and the dynamic created by this is the basis of its development and establishment, its ability to create a more or less extended area of influence. Rivers are the foundations of a city, then, of its layout, its transformations and its outward projection and one of the elements which best symbolise it and represent it and re-elaborate it in its various forms.

We could almost say that the relationship between cities and their citizens with their rivers is a nearness-distance fluctuation.

Water courses are not always a 'showcasing' front with which cities express their identities and best sides.

Rivers are also a threat, flooding cities and causing damage, destruction and death. When river banks are unhealthy and dangerous, cities move away, reject and fence them off and build embankments around them.

And in cases such as these, rivers attract another city, its less acceptable face, factories, mills, workshops and undesirable, malodorous and polluting installations, and also its marginalised city, those who can find nowhere else to settle and thus build their illegal huts and vegetable gardens along the river banks.

At this point cities separate off from rivers and push them aside, until – in a quasi nostalgic reaction – they try to return to them, to take possession of them once again. But they find them changed, transformed, worsened, polluted and made nasty. And so they attempt to tame them, adopt them and repacify them, to redesign their banks.

It is a cyclical story, this one, between cities and rivers and one which does not always have a happy ending but has great design and perspective value.

This dynamic is different in each single town.

In Turin, for example, with its four rivers cutting through the city, there are four different scenarios which are very different from one another: the Po, Sangone, Dora Riparia and Stura di Lanzo. Rivers pass through urban districts in very different ways, taking on conflicting identities.

In this sense building along rivers conjures up the environment and the landscape.

And architecture – more than ever in recent years – responds in very varied ways to these themes. In these contexts the task of architecture – and not just that with a capital A – is not simply to make the world more beautiful but above all to help man *live on the earth*, to carve out the best spaces and trajectories for his daily activities – living, studying, working and (why not?) also having fun, relaxing and spending free time pleasantly, getting lost in the forests.

And it is this, I believe, which should be the starting point of projects in these areas of the city: they should be spaces to live in, spend time in and dwell in.

7.2 The Project Towards the Landscape

I think of the sequence of insights at the heart of the considerations I am about to develop as quasi circular in the sense that, inevitably, when as architects we reflect on the relationship between architecture project and landscape, we must necessarily reason in relation to the urban degradation of the world we live in.

I believe I am saying nothing new in recognising that architectural design is in crisis and that this fact casts doubt on a series of values intimately tied to professional (and teaching) activities.

I have asked myself many times over the years what the deeper meaning of our trade is for man, in 'living on the earth' – to cite Heidegger (1976) – in the landscape, in the world. Those of us working in universities have sought answers, sometimes by moving away from assumptions and preconceived ideas and at others by seeking to dig down deeper in a quasi psychoanalytical way into the processes which every design adventure brings with it.

It is a tension towards doing architecture which often means tackling new themes, working on new projects, building new landscapes and new stories, feeding and reinforcing one's critical conscience, trying to find a more ethical approach to modifying places to transform them and make them into places in which it is possible to 'live poetically and full of merit'.[1]

[1] '[…] my intention was to underline [Hölderlin's verse] «full of merit, but poetically, man lives on this earth» not with Calvinistic ostentation of the weight of this grow-build logic but, to the contrary, highlight how our necessary work («lavourer», in Piedmontese, to plough, open up the earth, in contrast to «travail»), our learning to inhabit and have people inhabit is meaningless, cannot generate reality, truth if not also and together with opening up, a sometimes tiring breaking through of the «harshness of the existing»', Isola (1986).

In recent years on the other hand – unfortunately – we have witnessed with a degree of impotence a progressive devaluing of the role of architecture (and architects), a process which I believe has lost its way and moved in the direction of commodifying the profession, and building design has now become a matter of the lowest bid and shortest time frame.

As we regularly and carefully leaf through magazines, follow blogs and read the newsletters which increasingly clog up our email inboxes, we cannot but note a progressive cultural impoverishment of architecture as a discipline in the face, alas, of few and mainly public context reactions.

Architecture, understood as a tool for the building of top quality public spaces and ever better places and landscapes to live in, is obscured by an increasingly common desire to dramaticise which generates, and seems almost obligatorily to trigger, loaded and self-referenced images lacking in content and soul.

This can prompt conflicting emotional states in us: it can depress us, scandalise us or leave us indifferent, but the question that those who have formed an opinion on the usefulness of this juxtaposition of (generally fleeting) *slogans* are left with denial alone, rejection as the only escape route, is still open.

What I mean is that these facts create a *hiatu*s within the deeper meaning of design and its bond with the land, inhabitants, history and landscape.

This impoverishment of thought, I feel, is slow but inexorable, and the same is true of the downscaling of architecture's positive charge as a driving force for change in the face of the inflation of material, hi-tech models which seem to serve exclusively image bound ends in terms of product, the culture of the fleeting and the market.

A powerful void exists today between the city's physical presence and its citizens, between the *urbs* and the *civita*s with a tangible crisis and consequent inadequacy of contemporary *polis* models.

We cannot be indifferent to this state of affairs. I personally feel disquiet and a little indignation.

We must resist the onward march of this uprooting and seek to start afresh from the meaning of an architectural project, of that *recherche patiente* which Le Corbusier spoke of (1960), analysing case by case, enquiring into the needs and potential which a specific place expresses *in nuce*, revisiting all the specific features of the local area, culture and history each and every time.

To respond adequately with the right tone, what is needed is silence, calm, care and the chance to rethink things and return to places; to assess even the smallest, often latent signs which are there hidden in the landscape, in filigree, with great dignity; and then also to throw everything open to debate, start afresh and listen once again.

Every single time, we must try to define landscape once again, to add something new to our academic research.

And in this sense I think that our relationship with the landscape must be 'cultivated' in the form of education and teaching just as is generally the case with arts or foreign languages. If we are not able to recognise landscape because we have not been brought up to read it, we can look at it for ever without succeeding in seeing it just as can occur with a work of art or an unfamiliar language.

Le Corbusier (1963) said: 'la clef c'est: regarder… regarder/observer/voir/imaginer/inventer/créer'. It is precisely in this sequence of verbs, which might at first glance seem to be synonyms (but they aren't), that the key to our profession is to be found: look, observe, see, imagine and invent. These are the actions which an architect must always put into practice everyday in every project.

I believe that these attitudes, this approach to the subject, are fundamentally important. The right 'attitude' to work is crucial.

We are often distracted by images (books and magazines but also videos and Internet sites) and by the influences these have on us. But the construction of an opinion on architecture as a profession should be the primary objective of each and every architect.

Until this happens what is important is to know history in the sense of acquiring a direct experience, drawing elements with which to comprehend the traditions of one's own trade from the architecture of the past.

History, as Gregotti (2008) said, 'is the irremovable terrain we walk on, on which our state is founded even if [luckily, n.d.a.] it tells us nothing of the direction we should travel'. We have to learn to interpret it.

I believe that in our relationship with history, there can (or must) be an explicit, voluntary reference to *tradition*: as a showcase, recollection, citation or melancholy, or as nostalgia which is sometimes ironic – why not? – it is not tradition, however, in the sense of uncritically reiterating, unimaginative copying. Quite the opposite, I mean recognising the permanence of the past and reworking it in a contemporary interpretation. One of the most intriguing and characteristic features of our cities is precisely their architectural, material and long-term spatial stratification, their history.

We must, however, 'learn to forget and forgive history', wrote Aimaro Isola (1993b) citing Ricoeur. That is, we must have a passionate attitude to our past, look after it, think of it with *pietas*, with a critical ability to reread and distinguish, in what it has been and has been passed down to us, what is of value and must be maintained and enhanced from what should be forgotten and perhaps even cancelled out.

7.3 Design Along the River

Rivers were once the central attraction and localisation factor in man's history and settlement patterns. Water and flow have been factors of economic development and progress in social living standards.

The city-river dynamic can be seen as a metaphor for the way in which the relationship between town centres and the environment have determined human settlements over time but also of the way in which urbanisation has mutated over the centuries and in which we inhabit, practice and think city spaces and the urban landscape, of the way in which, in a few words, we experiment the urban.

In cities water courses have great symbolic value: from urban façade to centre of life and productive activities, place of *loisir* or, quite the contrary, factory hinterland and waste disposal, simply forgotten by its inhabitants and now a landscape scenario or place in which to recover a meaningful relationship with the environment and also human well-being, with the good life, with an awareness of the quality of life and the beauty of the city.

Rivers also have fundamentally important environmental functions and constitute an interweaving of relationships which cities maintain with their hinterlands. Dealing with such a complex situation requires great care and the tools used to manage it must necessarily go beyond the purely technical which are generally linked to the concept of 'green' understood in functional or environmental or hydraulic engineering terms or in relation to other subjects which share reference points with it.

Over time, and especially over the last century of intense urban development, the river-city balance has been destroyed, and intensely critical elements have come to the fore. Water is increasingly polluted as a result of only partially purified civil and industrial waste discharge; the management of high water has become more and more difficult as a result of excessive and often indiscriminate urban development (and thus impermeabilisation) of the soil precisely in those areas which once acted as a flood plain; or it has become critical as a result of hydraulic action designed to modify, straighten, channel and pipe normal river flow and expansion in an entirely unnatural way.

Land damaged by illegal building often cannot withstand the stress of climate change, and the consequent expense involved in putting increasingly frequent natural disaster damage right (the so-called water bombs) is often in actual fact much higher than the investment needed to prevent such disasters.

In many cases rivers have been transformed, become alien and gone from friend to foe. They are things to conceal, cover up, marginalise and to distance from the city's daily and social life.

The two European directives on water[2] and defence against flooding[3] would seem to pave the way, from a legal point of view, for the recovery of a fertile partnership between river and city. On one hand, it is important to pay attention to the quality of the water flowing through our cities. It needs cleaning and reclaiming to give cities back their rivers so that they can once again play a part in city social life as environmental air conditioners, bearers of well-being and beauty. On the other hand, we also need to restore safety and do something to ensure that citizens reacquire their faith in urban rivers. Rivers which instil fear in us every time it rains are enemies. We must ensure that floods are the exceptional events they once were, preventable and avoidable events which can also be faced when they occur with regular checks and planned management of potential flooding.

In addition to water quality and hydrogeological imbalance and soil erosion – certainly the pre-eminent themes from a strictly environmental point of view – we must also pay attention to the *urban regeneration* of river banks in particular in the most highly developed areas. Green area space design, the search for high quality

[2] European Parliament and Council directive 2000/60/EC dating to 23rd October 2000 which implemented a common framework of action on water.

[3] European Parliament and Council directive 2007/60/EC dating to 23rd October 2007 relating to assessment and management of flood risk.

standards to be respected in building and the most avant-garde technological solutions to reduce wasting energy and natural resources – these are the emerging city design priorities. And this must be accompanied by a serious rethinking of living spaces, of urban quality. These are considerations which must start by *listening* to places (Portoghesi 1989) and the ability to find a place in the dialogue between people and things which is peculiar to meaningful landscapes.

Especially in these recent new millennium years, the world has changed. Nothing authentic is left, perhaps just the skies and the geography of places. All the rest is ever-changing landscape.

So what is it? What does landscape mean to man today?

I believe that it is not simply what we see but also, perhaps, a combination of our points of view on what is around us, a sign of our perspective on things and a design of how we would like them to be.

To understand and plan the landscape, we need to immerse ourselves in it and, at the same time, to see, to imagine, how it could be, how we would like it to be.

The landscape is not simply a backdrop. It is neither photograph nor portrait, the scenery of our daily lives, but an entity, a reworked image of the memory of sensations linked to place or via the images of a film or the interpretation of the pages of a novel which describes it,[4] which encompasses all our senses.

Landscape can take various tangible forms. Each of us may have subjective perceptions relating to time or the play of light or our state of mind. Mountainous outlines or large walls can sometimes appear harsh; at other times, a curved roof can evoke a hilly horizon; a glass surface can reflect surfaces and transparencies, light and shadow; and a portico can frame an urban view.

In a certain sense landscapes are part of our approach to life, we have to inhabit them (and thus design them).

I believe that architecture has a *design attitude* to offer in response to this desire to inhabit, the ability to put forward synthetic images which interpret both the nature and deeper character of these places and are already written in their stones, which evoke a deeper awareness but also juxtapose continual gaps, shifts and alterations to the banality of the existing and direct us away from what is already there in the direction of what might be.

What I mean is that, in urban transformation projects, we architects should perhaps try to better define the sphere and limits of our action, interpreting the resilience of the city's critical places understood as the ability to autonomously and spontaneously reorganise, like the urban system's ability to resist, adapt and respond positively with innovative forms and techniques to the stresses and strains which changes (climatic, historical, economic and social) subject it to.

What if we were to try not so much to position our buildings on a site or to camouflage them but rather to design houses and places which are welcoming, to

[4] For Piedmontese like me, for example, the landscapes described in the novels of Cesare Pavese or Beppe Fenoglio are much more than a simple literary description. They often represent an approach to life, an attitude to others, an emotion which can escape pure observation of place.

host those living in them – in the meaning Jabés[5] gives to this term – and not simply to contain them? What if we were to succeed once and for all in convincing ourselves that the threats we often attempt to defend ourselves could actually be opportunities? What if we could understand the potential these very threats might offer to cities, areas, nature? We would be able, I believe, to bring new languages into the equation, make them richer, go beyond what has already been done, beyond what we know.

The city project, or rather a project for the city, should be capable of defining a strategy, putting forward new frameworks and implementing the urban system by getting all local players involved in it (from utilities to universities, from citizens to the service society) and garnering all the necessary skills.

Today we need to know how to read the differences which make up our know-how carefully and take on board the inhabiting requirements which emerge to give a responsible, authentic, well thought out, i.e. *scientific* response, to them.

To give such a response, we need to be able to muster our however grossly limited know-how, as well as that gained outside our own discipline and above all that consolidated in the however tormented everyday practices of each of us.

The design attitude I referred to earlier is not exclusive to architecture. It is multidisciplinary; it is an attitude which architecture and town planning shares with a geography and history which can interpret the universe, instead of simply reducing it to objectivising paradigms of descriptions, data and documents. And many others disciplines are capable of this attitude: technology, the sciences of building and energy, environmental engineering and ICT.

I am certainly not saying that we should abandon that architecture *arché* which has built up over history – this remains the filigree of all our work – but I believe that we should take such differences on board in the wealth of fluid and varied perspectives which the inhabiting requirement poses us.

This means *opening up* the language of our disciplines to other forms of knowledge, other cultures and other people.

A multiplicity of explorations, on the diverse project scales and according to different points of view, which must have in common this *design attitude* aimed at making the project's places hospitable, to make them lived in. It is a way of life which leads us to immerse ourselves in places in a self-aware way and perhaps with the humility of looking at architecture from the point of view of those living in it, of designing the landscape from inside too, so that buildings once again dialogue with those living in them, belong to the collective imagination of those using them, contribute to improving environmental quality.

There is a very beautiful painting by Caspar David Friedrich (Fig. 7.1) which depicts a landscape in the fog seen from a hill. The painter, however, added a person (himself?) to the painting (and thus to the landscape) seen from the back looking

[5] 'On this side of responsibility, there is solidarity. On that side there is hospitality. Giving in to the needs of hospitality, to its unspoken requirements, means to some extent grasping the practice involved in our dependence on others', E. Jabés (1991).

Fig. 7.1 Caspar David Friedrich *Wanderer above the sea of fog*, 1818 (Source: Hamburg Kunsthalle collection)

towards the horizon. Now I believe that like this person, we must try to 'enter the landscape', to take our place within it and not simply look at it from outside, from beyond the frame. Designing a landscape, I believe, means removing the picture frame, opening the window and understanding that we too are in the landscape we are designing.

I have already said that every architectural project by its very nature generates change. The difficulty consists in ensuring that this change is for the better.

The first concern should, then, be trying not to damage sites. We need to think how buildings can dialogue with what was there before them in an interesting, intriguing way.

Modifying the landscape sometimes means making it denser, building in any gaps that still exist. But building does not always mean pouring tons of concrete onto the land. Sometimes it simply means creating earth movements or designing a cliff of stones or a row of trees, working on perspective just as, on the other hand, landscapes are not always that uncontaminated nature so dear to intransigent environmentalists, and in fact these hardly exist anywhere today.

7.4 Urban Regeneration and Resilience

The invention of new landscapes thus becomes an urban regeneration project in which the resilience issue takes on new meanings and becomes an entirely new specific discipline. In physics resilience is a material's capacity to absorb shocks without breaking, while in psychology it is the human ability to positively reorganise one's life when faced with difficulty without damaging one's own identity. In architecture and also in town planning, resilience is 'a specific idea of intelligence capable of remodelling itself around the complexity of events which are destructuring the cities' (Infante 2013).

Dealing with what remains of past times does not necessarily mean going in search of the traces and signs of an often compromised image so much as attempting to 'sew up' the loose ends (the leftovers or tatters). It is certainly an arduous and challenging task which cannot and must not take the form simply of grand infrastructural work (as it often did in the past) but quite the contrary involves the building of a capillary network of relations some of which are simply virtual, informal and not necessarily physical networks whose purpose is to give new meaning to the city's places and engrave them into the city's consciousness.

It is thus a weak but widely disseminated reconnection to be implemented, perhaps, beyond the confines of scientific rigour, geometry, rules with a preference for a slightly out-of-focus thought process which accords better with the reality of unconsolidated, degraded spaces, of the dimension between open space and built space, between the city's masses and the energy of those who live in it.

It is certainly important to pay attention to the real historic and environmental values of places, but designing in these borderline landscapes also means having the courage and sometimes taking the responsibility for turning an area's framework on its head and establishing new spatial and hierarchical relationships on condition that the purpose of these new developments is always to improve urban and environmental quality with the ultimate goal of diffused quality.

The difficulty involved in regaining the past is not the reproposing of historic materials or reusing lost technologies. These are all accessible in modern building sites. This has already been done. The difficulty lies in a concept of inhabiting which has changed entirely and this in the space of just a few generations. Sixty years ago we were still building out of need with poor and few technologies confirmed by lengthy experience and the solid reference points of schools and managerial or aristocratic classes. It was all very straightforward. Learning was a challenge but the direction to take or beliefs on the forms to be used weren't.

Today, on the other hand, we build for bogus reasons and fleeting fashions with an avalanche of technological data and a plethora of ever-changing laws in the total absence of formal reference points.

On the other hand, architecture's 'star system' cannot prompt virtuous behaviour but is more likely to trigger petty jealousies and unhealthy emulation ambitions.

How can we transform and recover river banks to make them accessible to the inhabitants of our era? How and why should we build once more in these places? It is architects who should be telling us this; it's their job.

I don't believe, however, that it is an easy task. Quite the opposite, I think it is an arduous task and one with no secure reference points. And this is even more the case today when we have to do daily battle with the Internet and thus with an infinite array of uncontrolled possibilities. Our times are characterised by an extremely and excessively broad range of possibilities and, above all, one which is not subject to criticism except as far as costs and practicability are concerned. This is too little for a value judgement.

In this context schools can perhaps help those studying architecture today to work with wisdom, competence and intelligence in such powerfully compromised landscapes (at least in Italy) but at the same time so packed with potential.

This hope has grown from an awareness that for many years now in our Politecnico di Torino school – in the Design Laboratory, in degree theses and workshops – we have been attempting to familiarise students with the issues of landscape and environment and urging them to put forward projects to redevelop abandoned areas.

Some of these ideas are very courageous because they are frequently developed in very short time frames (a few weeks) and perhaps, above all, because they are the work of young students who still haven't learnt the tricks of the trade fully, who haven't been compromised by professional practice. In any event, however, they have the great merit of seeing outside the box, of having pushed at the boundaries a little, to design new landscapes which are innovative in formal terms but also in the use of materials and building technologies.

They are often ideas which have grown from a shared conviction: the intrinsic beauty of the landscape, its evocative power, its capacity to become the source of inspiration for a project.

But every student gives the subject his or her own personal touch and interpretation, and almost all design ideas are valid, plausible and practicable, in demonstration of the fact that there isn't a single way of doing things; there are no preconceived formulas, no a priori right or wrong.

As projects grow out of places, they are strongly rooted to the soil they rest on, belong to the landscapes they develop and accompany their transformations.

As Luisa Bonesio (2001) has affirmed, every place, every landscape, has its own specific identity and physiognomy:

> there is no single solution applicable to any place or culture but neither is there a single space for subjective *creative* judgement. A place's dimension, its *genius loci*, dictates implicit rules which, we can argue, have been respected when the result is that of a good form and a profound and stable harmony which does nothing to disfigure the place's physiognomy but makes it recognisable in every intervention.

It is not by chance that the way of we build (and dwell) in our valleys, countryside and hills has remained unchanged for centuries and has been passed down to us. These building methods were and still are today the most suitable to an environmental context characterised by especially harsh and challenging living and working conditions.

This certainly does not mean that nothing new can be invented, that it has all already been said and that all research is now vain.

But all new styles must use what already exists as the starting point, from an awareness that the solutions adopted by our ancestors were chosen for a reason and have stood the test of time precisely because these reasons were good ones, had a solid basis and were rooted in the territory and understood as space, environment, soil, climate, culture and so on.

We can certainly replace solid chestnut wood with wood laminate, stone walls with reinforced concrete and larch wood planks with sheet metal and then create new forms on our computers: polylines, *nurbs* and *bloboidali* volumes have no secrets for architects of our day. New forms and technologies, however, must face up to the world around them and find a harmony with it, *a sense of proportion*.[6] They must be capable of adapting to their locations gracefully.

This is the point: finding an equilibrium between what exists and what we want to do, working with similarities or differences but always seeking to modify in the sense of improving even of revolutionising, as I said earlier, but never turning upside down.

7.5 Conclusion

To rebuild and regenerate our landscapes, give them fresh meaning, breathe new life into them and bring people back into them; we need to start from here, from the ethical design issue: the quality and beauty of a building depends to a considerable extent on the *habitability* of the site to be built on.

Improving places means making them hospitable, and architecture, to be hospitable, must integrate into a site elegantly and gracefully. To make our presence felt in the dialogue with the pre-existing, we don't need to scream and shout. We need to look, observe, listen, understand and interpret.

A complete reading of the theme, a precise reference to users, in such places, can help us in our approach to the project. We must take up a site's culture and history.

Landscape identities require safeguarding, enhancing and regaining but not crystallising or mummifying. Landscapes evolve and mutate too; they transform together with the civilisations living in them. It is fundamentally important, however, to manage this evolution and provide solutions to the various demands made on them (coming from the economic, political, social and other worlds) which are not evasive or shallow but capable of adapting intelligently to the changes we are experiencing.

[6] 'The question we pose today is finding a sense of proportion which allows the complexity of the world to be preserved in both time and space terms without us getting lost in it', Berque (1995).

This is our responsibility (from Latin *responsare*, that means to give a respond to) as architects and I believe that this is the direction our projects and styles should move in.

References

Berque A (1995) Les raisons du paysage. De la Chine antique aux environnements de synthèse. Hazan, Paris

Bonesio L (2001) Geofilosofia del paesaggio. Mimesis, Milan

Bruna F, Mellano P (2006) Architetture nel paesaggio. Skira, Milan

Bruna F, Mellano P (2011) Ancora e sempre il paesaggio. In: Fagioli M (ed) Dove va l'architettura? Aion, Florence, pp 22–25

Gregotti V (2008) Contro la fine dell'architettura. Einaudi, Turin

Heidegger M (1976) Building dwelling thinking. Italian edition: Vattimo G. Costruire abitare pensare. Mursia, Milan

Infante C (2013) (Ri)generazioni urbane. La nuova ecologia 5:48–49

Isola A (1986) Il brutto e la periferia. In: Bazzanella L, Giammarco C (eds) Progettare le periferie. Celid, Turin

Isola A (1991) Per un'etica del paesaggio. Atti e Rassegna Tecnica della Società degli Ingegneri e degli Architetti in Torino, 56, May–June, Turin

Isola A (1993a) Necessità di architettura. Atti e Rassegna Tecnica della Società degli Ingegneri e degli Architetti in Torino, 2, September, Turin

Isola A (1993b) Pensare il limite, abitare il limite. In: Giammarco C, Isola A (eds) Disegnare le periferie. NIS, Rome

Isola A (1995) Abitare il paesaggio: uno sguardo dal nulla. Atti e Rassegna Tecnica della Società degli Ingegneri e degli Architetti in Torino, 2, September, Turin

Jabés E (1991) Il libro dell'ospitalità. Raffaello Cortina Editore, Milan

Le Corbusier (1960) L'Atelier de la recherche patiente. Vincent Fréal, Paris

Le Corbusier (1963) Carnet T70, n. 1038, 15/08/1963. Casabella 531–532/1987

Morris W (1881) Prospects of architecture in civilization (delivered at the London Institution, 10/03/1880). In: Morris W (ed) Hopes and fears for Art. Longmans, Green and Co. American Editions, New York

Portoghesi P (1989) L'architettura dell'ascolto. In: Ciucci G (ed) L'architettura italiana oggi – Racconto di una generazione. Laterza, Bari

Chapter 8
A Future for Our Past: Textile Colonies in Catalonia

Joaquín Sabaté Bel

Abstract Many regions' industrial history has involved the creation of factory towns, but in Catalonia, their proliferation and development shaped the physical and cultural landscape in a unique way that has entailed a specific set of problems and opportunities today. These factory towns, or "colonies," are located in rural areas along the rivers that run from the Pyrenees to the Mediterranean; they grew in isolation from each other and from existing centers of population, beginning in the second half of the nineteenth century. However, the colonies did not themselves grow into diverse urban hubs, but instead remained small and self-contained developments, with no residential or commercial activity unrelated to the factory. When the factories closed during the second half of the twentieth century, these colonies entered a phase of crisis, and many residents moved away; today the colonies are deteriorating and have been largely abandoned, with little industry or commerce. In recent years, we have worked together with government and residents to develop master plans for the colonies in order to improve quality of life and to protect the heritage they constitute, converting them into an asset for the region's development.

Keywords Industrial colonies • Heritage park • Cultural landscape

8.1 Introduction

Textile colonies, or factory towns, are one of the most characteristic phenomena of Catalan industrialization. They implied a quite unique economic, industrial, and social model, becoming singular features in our river landscape. Between 1850 and 1880, hundreds of factories and worker's dwellings were built. Soon colonies

J. Sabaté Bel (✉)
Urbanism and Regional Design, Polytechnic University of Catalonia,
Calle Jordi Girona, 1-3, 08034 Barcelona, Spain
e-mail: joaquin.sabate@upc.edu

R. Ingaramo, A. Voghera (eds.), *Topics and Methods for Urban and Landscape Design*, Urban and Landscape Perspectives 19,
DOI 10.1007/978-3-319-51535-9_8

Fig. 8.1 Heritage resources: canal houses

became small factory towns, with a small school; always a church; housing for workers and their families, the owner's residence, and the one of the factory director; gardens and orchards; and sometimes even a small theater.

The irregular flow of Catalan rivers was counterbalanced by the existence of many natural water jumps, cheap infrastructure, abundant labor force, and quite favorable legal conditions. Along the Llobregat River, one may find dams every 10 m of slope, and along the Ter River, the length of the canals is equivalent to the length of the river itself. Railways and roads followed, and over 50 of these textile colonies grew well into the twentieth century.

The colonies represent an industrial and social project that profoundly changed the region's natural and agricultural landscapes, and they thrived as towns while the factories continued to provide employment. In the second half of the twentieth century, though, nearly all the colonies' factories closed; the colony represented an outdated mode of production in the modern age. Limitation of energy and cost of transport, social conflicts, international competition, and offshoring led to a growing crisis. This crisis was so strong that today we hardly find operative factories in our colonies. Many of them disappeared dramatically due to their productive obsolescence, to the little residential comfort, and to the lack of services.

But some of them are still alive as small villages, and they have consolidated a rich heritage of dams and sluices, canal houses, canals, factories, chimneys, different kind of dwellings, public facilities, and gardens (Figs. 8.1, 8.2, and 8.3).

The Catalan colonies are an impressive piece of cultural heritage. They are also remarkable as examples of dense urbanization in otherwise rural areas (Fig. 8.4).

Fig. 8.2 Heritage resources: canals

Fig. 8.3 Heritage resources: canals

Fig. 8.4 Cal Prat

But along the Llobregat River, they are unique not for the high quality of their design but for their remarkable number and concentration. The abundance of colonies with a clear relationship to each other allows them to be understood together as an industrial landscape unlike any other.

At the beginning of the twenty-first century, after 150 years of productive history, Catalan textile colonies were affected by their strongest crisis. At this stage, the work of local stakeholders (grassroots) and scholars was quite important in order to recognize and appreciate the value of the colonies. The Catalan Science and Technic Museum began to document and collect artifacts, and catalogues of the colonies' architectural patrimony were prepared as a step toward their preservation. In the late 1990s, the Government of the Province of Barcelona commissioned a group from the Catalonian Polytechnic University (UPC) Department of Urbanism and Regional Planning to produce a study of a section of the Llobregat River.[1]

Soon afterward the Regional Government commissioned again our group to develop a master plan, with the aim of promoting and coordinating different efforts

[1] Heritage Resources Inventory and Regional Structure along the Llobregat River. Province Government of Barcelona, 1999–2001.

and interventions trying to regenerate this impressive heritage.[2] Instead of viewing the colonies merely as obsolete and decaying artifacts whose physical elements or structure must be preserved, these plans instead interpreted the system of colonies as an economic and social resource that can be exploited for the future benefit of its residents and the regional economy.

8.2 A Master Plan to Transform Industrial Colonies Along the Llobregat River

The Llobregat River Industrial Colonies Master Plan defines a geographical unit of around 30 km with a rich sequence of 18 colonies, embedded in a narrow river valley, and affecting 2000 inhabitants. A quite small factory town that carefully fills all the useful space between the river and the old path occupies every useful meander. Such a narrow valley always conditioned communications. A currently abandoned railway line and a traditional road ran along the watershed connecting villages and colonies. The river landscape is quite discrete with little useful land and a modest riparian forest. But it contains a rich cultural heritage of colonies, farmhouses, and chapels.

Today the inhabitants of these colonies may be considered second-class citizens. They don't have public transport, neither the most basic facilities. The factories closed and residents abandoned progressively the colonies, contributing to their aging structure.

So we arrived to the conclusion that colonies' regeneration requires their economic promotion. The capital gains have to help improving open spaces and paying the extra cost of recovering old buildings. An active colony requires an active factory. This means occupying again the factories with compatible activities. It requires promoting high added-value uses and even facilitating the change from industrial to residential or tertiary activities. Therefore, the main goals become to impulse cultural tourism, to promote a qualified productive activity, and to preserve natural and cultural heritage. We should try to find an accurate balance between conservation and modernization and an agreement between public and private interests that encourages effective preservation, linking urban improvement to economic promotion, particularly of those sectors that may generate profit. The motto could be preservation through transformation.

It is important to find the role of each colony within the broader network in the river valley, to outline strategies for attracting economic activity, including the creation of a heritage park to promote tourism and local communities' self-esteem. By highlighting the colonies' cultural assets and their historical importance, they

[2] "Pla director urbanístic de les colònies del Llobregat," approved by the Regional Government on 8 June 2007 and published in *Diari Oficial de la Generalitat de Catalunya* no. 4940 (3 August 2007), exp. 2005/18060/B.

Fig. 8.5 L'Ametlla de Merola

may be able to be transformed from blighted and isolated outposts to connected and relevant urban centers.

The master plan provides also a level of official recognition and a path for future growth that is slowly attracting private economic growth in the colonies. At a territorial scale, it promotes links between colonies. Four of them become main entrance doors of a proposed heritage park. The narrative is crucial, and therefore these doors are called to play a role as a cultural entrance, a museum colony, a center of the park, and a commercial entrance. The river path has been refurbished, cleaning it and completing some tracks with small bridges. The old territorial road has been recovered as a local route with a bike trail.

At an urban scale, colonies constitute an active and complex heritage. Its main value lies in the relationship between buildings, open spaces, and the river landscape (Fig. 8.5). Current protection criteria are too specific and heterogeneous, sometimes limiting too much or, on the contrary, admitting demolition or disrespectful additions. Therefore, we draw accurate morphological surveys and we also researched the historical origin of each colony. A complete catalogue of buildings, plazas, paths, gardens, forests, and orchards was elaborated in order to identify and safeguard their structure and their genetic code.

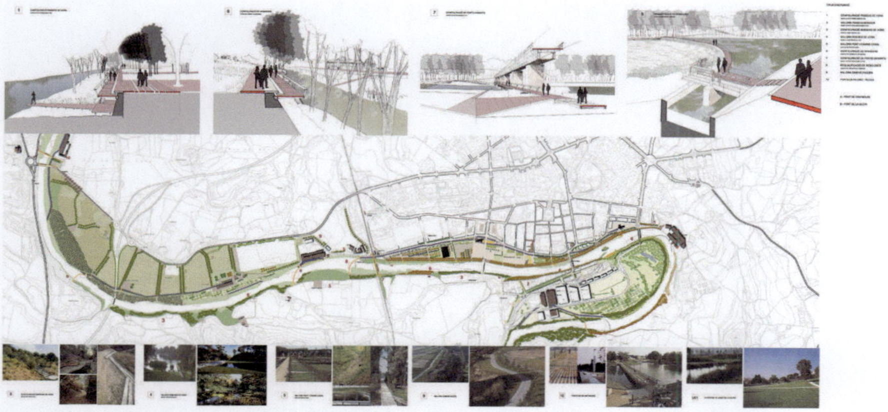

Fig. 8.6 Interventions along the river

One of the main goals is to convert old colonies into livable neighborhoods. It tries to solve the colonies' many deficiencies in terms of access to public services and infrastructure. In addition to modernizing telecommunication networks and technical infrastructure, it proposes better connections to each other and nearby towns via public transportation and improvement of roadways and footpaths. These, too, are necessary for attracting business, but they are also important for improving residents' quality of life. By carefully managing the colonies' future changes and growth, the plans also attempt to allow for a positive and respectful relationship between the industrial and residential aspects of the colony.

But a clear requirement is also achieving a minimum density to guarantee basic services and establishing an optimal growth according to the capacity of each colony. Therefore, the master plan establishes a gross density of 50 dwellings per hectare and a minimum of 100 dwellings for each colony. Potential growth is limited to one fourth of the actual dwellings and conditioned to previous rehabilitation of the existing ones. New constructions should be devoted to social housing with priority for inhabitants of the area. In order to face these growth compositional criteria, building typologies and diverse residential programs, from semidetached dwellings to row houses or small blocks, are proposed.

The basic aim is to maintain the morphological patterns through substitution or completion of the urban fabric. Therefore, special attention is paid to eliminating additions that are not consistent with the identity of each colony. For each one of the colonies and the surrounding territory, tailored interventions, some of them quite modest, were proposed, for example, how to improve natural flora and the needed steps to do it. And this research helps defining the treatment of riparian vegetation. It helps also conditioning some sections of the civic trail (Fig. 8.6).

The master plan proposes some other interventions along the river. The river offers rich opportunities to strengthen its quality, improving some itineraries along it: stressing its scenic values, improving traditional paths, or the embankments, and

Fig. 8.7 Interpretation center in an old railway station

fixing small bridges that cross it or natural balconies that let people enjoy the landscape. It also proposes to use existing infrastructure to hang pedestrian gateways, improving the access to the river or anticipating meeting points along the routes.

Some of these designs have been already developed, like an interpretation center in an old railway station (Fig. 8.7) or recovering one of the owner's houses to host public facilities or improving the access to the different colonies and recovering old textile factories as restaurants, or start-up's incubators. In some of the colonies, the residential fabric begins to be completed.

And colonies acquire with all these interventions a new residential character, but they also recover economic activity and enrich their functional complexity. This implies a respectful coexistence between a livable colony and an active factory. The colonies are currently in a liminal phase, as they transition from an obsolete production model to a new life as modern urban centers and as landmarks and nodes in a broader cultural landscape. In order to successfully regenerate the industrial colonies and their region, economic activity must return to the factory towns. The colonies' future transformation may entail growth or modification. So we have to understand this and not seek to impede development, but rather to guide it in a manner respectful of the colonies' heritage. Through their transformation, the colonies are being preserved.

Within a heritage park, the colonies provide a guiding theme, but they also relate to a broader narrative that unifies the industrial history of the entire river corridor. In studying the Llobregat axis, we broke the course of the river down into smaller chapters that together form a legible narrative of the region's industrial progress and its interaction with the landscape. From the river's source to the sea, these include cement factories and coal mines in the Pyrenees (c. 1900–1920), railways between the mountains and the plains (c. 1930), the Baells reservoir and the hydroelectric dam that controls the river's flow (1970s), the industrial colonies of the River Park (c. 1860 onward), a medieval irrigation canal (fourteenth century), the monastery of Montserrat and its various forms of access for the tourist industry (railway, funicular, cable car, twentieth century), and the irrigated farmland of the man-made delta near Barcelona (nineteenth century) that today is part of the Llobregat Agrarian Park. These chapters can of course be read chronologically, as well, to explain Catalonia's industrial development from medieval canal systems to factory towns to modern dams and the tourist industry, recently a bright spot in otherwise slow economic times.

8.3 Lessons from Heritage Park Projects

The growing number of projects for defining cultural landscapes and the recognition of their potential as catalysts of regional economic development have led us, together with some colleagues of the MIT, to study many examples. Our aim has been to learn from these experiences and to apply our understanding of cultural landscapes to the design of a concrete river area in Catalonia that we have labeled the Llobregat River Heritage Axis.[3] Some conclusions come from our analysis, in which we paid special attention to the concepts, methods, and tools we found most useful and broadly applicable. Departing from the most common and relevant aspects, and bearing in mind the need to remain sensitive to the unique situation of every project, we proposed a decalogue of lessons to be kept in mind in any cultural landscape project.[4]

[3]VV.AA. Designing the Llobregat Corridor. Cultural Landscape and Regional Development. Catalonian Polytechnic University and Massachusetts Institute of Technology. Barcelona, 2001.

Sabaté, J. "Paisajes culturales. El patrimonio como recurso básico para un nuevo tipo de desarrollo" en Urban n° 9 (pp. 8–29). ISSN 1138–0810. Madrid, primavera 2004.

Sabaté, J. "Paisajes culturales en Cataluña: un reto de futuro" en Espais n° 50. Generalitat de Catalunya. Barcelona, 2005.

Sabaté, J. "De la preservación del patrimonio a la ordenación del paisaje" en Urbano, año 7, n° 10 (pag. 42–49). Universidad del Bio-Bio. Concepción (Chile). ISSN: 0117–3997.4

[4]See Gestión del paisaje. Manual de protección, gestión y ordenación del paisaje. Ariel, Barcelona, pp. 625–642, and La construcción del paisatge de les colònies. Una aproximación morfològica. In: Colònies industrials. Departament de Cultura i Mitjans de Comunicació, Generalitat de Catalunya, Angle Editorial. Barcelona, pp. 114–131.

8.3.1 Every Heritage Park Must Tell a Precise Story

All the most successful projects have been based around conveying a cultural or historical narrative, not necessarily monolithic, but legible. It is important to bear in mind that not just one story exists. But in order to best educate visitors and preserve local identity, it is best to adhere to a specific and comprehensible interpretation, which should generally be that which is coherent with the available resources. These story and interpretation are essential for tying together distant landmarks so that they relate to each other and reinforce one another and to orient tourists, scholars, and visitors throughout the experience with respect to a general outline. These resources thus acquire a greater regional significance through their connectivity.

8.3.2 All the Involved Agents Should Discuss Intervention Goals

The main goal shared by all the studied examples is the integration of preservation with education, recreation, tourism, and economic development. In most cases, this implies a tight collaboration between different political administrations, local or regional institutions, and interested private agents. Cooperation is crucial for the success of a heritage park; shared goals must be defined to ensure that all parties are on the same page.

8.3.3 Coherent Boundaries and Itineraries Must Be Defined

One of the first measures to be taken is the precise and reasoned delimitation of the territory to be planned, with special attention to its resources and history, to its uniqueness, to those characteristics that merit preservation, and to its interpretation. But sometimes the area is so broad, so rich in resources, and so diverse in its identity that different narratives within it must be distinguished. In those cases, it becomes useful to divide the area, to define the different messages, and to attribute a role in their common history to each section. Themes are generally chosen so that these landscapes can be "read" like chapters of a book. Each chapter must have its own distinct identity, while contributing to the overall narrative. Chronological ordering can be a useful tool, where possible; competition or superposition of narrative segments should be avoided. Quite often the decline of the territory itself is included with a clear pedagogical intention, highlighting the potential of the heritage park as a means for its reactivation.

8.3.4 The Journey Is Critical

A typical – even essential – strategy is to design itineraries that link the heritage park's landmarks together. The actual experience of the journey through the landscape, guided by the heritage park's narrative arc, is an essential characteristic. It is important to define routes that can be followed on foot, bike, car, and public transportation. In some cases, this itinerary is enhanced by period transportation (horse, trolley, boat). The journey is not only important for conveying suggested narratives to explain the history and the present culture of the landscape but also for creating the heritage park's image. This image can be useful for drawing visitors and for improving the region's self-esteem, improving the general perception of the area. This perception is hardly insignificant; rather, it is a crucial part of the revitalization of the regions in question.

8.3.5 The Story Must Be Rigorously Documented

The narrative should be original and coherent with the available resources and, most important, well documented. Therefore, many of these projects begin with a complete inventory of resources: the basic ingredients of the story, the foundation of the interpretative process, and the principal attraction for potential visitors. In the elaboration of these inventories, community members should play a fundamental role, through meetings of the entire interested population and by working with experts and academics. Initiatives must be rooted in local communities.

8.3.6 Residents Themselves Are an Important Cultural Resource

The residents are essential to the future of a heritage park, due to their knowledge, their memories, their role in the narrative, and frequently, their enthusiasm. They are, in the end, the true reason for launching such an initiative and the most interested agents in reappraising their own heritage. As soon as the community's self-esteem is reassured, residents no longer feel like they belong to a region in crisis and that they can begin to build a new future from the heritage landmarks all around them. The best initiatives recognize the value of the local community and incorporate residents into heritage parks' promotion and design. The best projects are widely participative. The most important aspect to launching a project is bolstering resident communities' self-esteem. Visitors, museums, and investment will come in time.

Memory is another basic cultural resource: the work done by anthropologists, sociologists, historians, geographers, and archivists is invaluable. When the remnants disappear, the collective memory, the shared heritage, and the cultural traditions of communities are as important, or even more so, than its monuments. We should thus lend special attention to the memories associated with specific cultural landmarks. We must compile histories and document traces of the past, before they are lost. For in-depth research on the history of an era, a society, the transformation of ways of life or of landmarks is a fundamental ingredient to the most compelling heritage park initiatives. Enhancing this aspect through classes, seminars, and publications at the heritage park itself makes for considerable added value.

8.3.7 The Most Successful Initiatives Come from the Bottom Up

Grassroots movements, made up of local residents who love their region and want to make the most of its landmarks, are frequently the most successful. Where possible, residents should be included in the planning and development of the heritage park; they understand their community's needs, and they are often the most passionate about their heritage. It is a difficult undertaking to run a successful heritage park in which no local groups or organizations play a relevant role.

8.3.8 Administrative Complexity Can Be an Advantage

Generally speaking, regional initiatives typically involve input from different levels of government and numerous actors, and a heritage park's administration implies a superimposition of government and bureaucratic responsibilities and relationships. Despite first impressions, complexity at the administrative level should not be seen as a disadvantage, as the involvement of many parties can create new opportunities and can provide more resources. Far from seeing this as a problem, we should regard administrative overlap as an opportunity to bolster combined efforts and make the most of a new participative culture. It is important, though, that all the groups work together toward shared goals.

Indeed, cultural landscapes are capable of bringing together vast regions and scales of activity, transcending the limits of a specific level of government. These can be unifying projects, and as such the inclusion of many actors can be quite positive.

8.3.9 *Official Recognition Is Generally More Important Than Economic Subsidies*

There are various types of recognition, typically entailing official protection or financial subsidy. This typically comes from government institutions, but among the most prestigious are UNESCO's designations. They are important not only for the obvious benefits they confer (investment or tax relief, preservation of monuments and landscape) but for the cachet that such official recognition carries. Often these official designations imply more obligations than financial resources, but their ability to attract increasing numbers of visitors (and eventually investment and jobs) outweighs any negative effects. They are also effective tools in improving community self-esteem.

8.3.10 *The Heritage Park Must Have a Clearly Defined Physical Structure*

Plans and projects for cultural landscapes are still a novelty. This has entailed the need to develop new methods and tools. But in almost every instance, we can see a basic structure with some repeated components:

1. The boundary and sub-boundaries of the park
2. The park's heritage resources and services
3. Access points, interpretative centers, and museums
4. The paths that link the above elements
5. The visual (and administrative) limits of the project

With a bit of license, we could compare these to the five components of the syntax proposed by Kevin Lynch in his book *The Image of the City*:

1. Regions
2. Landmarks
3. Nodes
4. Itineraries
5. Edges

Like Lynch, we can demand that these elements meet certain requirements in order to ensure greater legibility and to strengthen the identity of the cultural landscape. Ideally, each of these elements should have the following formal characteristics: singularity, formal simplicity, continuity, relevance, clarity, directional differentiation, visual reach, and names related to meanings.

8.4 Changing Landscapes: Toward a New Planning Paradigm

Cultural landscapes and heritage parks show a rising importance in locally based development. But we should not consider this as the final goal. Twentieth-century regional plans were mainly concerned with population dynamics and industrial development, providing housing and locating activity poles. Zoning and layout of big infrastructure became fundamental tools.

Today some of the most outstanding recent plans attend to a new dichotomy: nature and culture, as parts of a unique concept, heritage. And cultural landscapes may become a useful vehicle to achieve the goal of constructing more diverse and identity-rich environments and of reactivating certain territories.

Today our concept of heritage extends far beyond the individual monument and can be understood on a territorial scale in the form of "cultural landscapes," a term that is hard to define but essentially connotes humanity's physical and semantic transformation of a site. As urbanism, too, has progressed to the territorial scale, it has sought to address such cultural landscapes, preserving them and incorporating them into its own process of reshaping the landscape. The "heritage park" has been a major instrument in urbanism's approach to cultural landscapes, entailing physical preservation, but also education, residents' participation, and economic development.

In Catalonia, heritage parks that have been developed around an agricultural river delta and around a series of textile factory towns have followed in this relatively recent tradition, seeking to preserve the region's heritage while also spurring its future development and attending to the needs of its present residents. Particularly in a postmodern age that simultaneously values and destroys authenticity, cultural landscapes are unique assets; it is urbanism's role to plan them and integrate them into future development.

Cultural heritage is not simply something to be preserved. We may find development opportunities upon territorial and historical values. We need to beat a strict conservationist position and learn to work with heritage resources within transformation processes. Current interventions may also generate future heritage. Preservation through transformation could be the synthetic statement.

In this sense, cultural landscapes may play an important role, because they constitute the expression of memory, of the identity of a region. And this identity offers the possibility to be enriched. It's not a matter of mere maintenance of a cultural legacy. Today more than ever, against the extension of the so-called non-places, toward many landscapes becoming banal ones, our interventions should seriously consider their identity and memory.

This would be one of my conclusions. The genetic code of each landscape shows its alternative. And in order to intervene in it, we must know it and respect it.

I would like to end by remembering what a great writer and a greater person, the Nobel Prize winner Jose Saramago, told us some years ago in the island of Lanzarote. The island authority wanted to enlarge an existing road that runs along a beautiful

Fig. 8.8 La Geria (Lanzarote)

testimony of human labor, called La Geria, and that would seriously impact it (Fig. 8.8). Many professionals were intending to stop it with different arguments and alternatives. But Saramago stated the most clear and powerful argument by saying "... a society that does not respect its territory, and the previous work footprint upon it, is a society that does not respect itself."

In the same sense, we should learn to respect our territories and therefore respect ourselves. The genetic code of every territory, the information it contains, should be always the basis of its future design.

References

Lynch K (1960) The image of the city. MIT Press, Cambridge

Sabaté J (2004) Paisajes culturales. El patrimonio como recurso básico para un nuevo tipo de desarrollo. In: Urban 9:8–29

Sabaté J (2005) Paisajes culturales en Cataluña: un reto de future. In: Espais 50. Generalitat de Catalunya, Barcelona

Sabaté J (2006a) De la preservación del patrimonio a la ordenación del paisaje. In: El paisaje y la gestión del territorio. Criterios paisajísticos en la ordenación del territorio y el urbanismo. Consorcio Universidad Internacional Menéndez y Pelayo de Barcelona, Diputación de Barcelona, Barcelona, pp 329–342

Sabaté J (2006b) Paisajes culturales en Cataluña: el eje patrimonial del río Llobregat. In: El paisaje y la gestión del territorio. Criterios paisajísticos en la ordenación del territorio y el urbanismo. Consorcio Universidad Internacional Menéndez y Pelayo de Barcelona, Diputación de Barcelona, Barcelona, pp 531–548

Sabaté J (2009) Proyecto de parque patrimonial fluvial del Ter. In: Gestión del paisaje. Manual de protección, gestión y ordenación del paisaje. Ariel, Barcelona, pp 625–642

Sabaté J, Vall P (2009) La construcción del paisatge de les colònies. Una aproximación morfològica. In: Colònies industrials. Departament de Cultura i Mitjans de Comunicació, Generalitat de Catalunya. Angle Editorial, Barcelona, pp 114–131

VV.AA (2001) Designing the Llobregat Corridor. Cultural landscape and regional development. Catalonian Polytechnic University and Massachusetts Institute of Technology, Barcelona

Chapter 9
Agro-Urban Models in the Sarno River Plain

Carlo Alessandro Manzo

Abstract The landscape of Sarno river plain, among Vesuvio and Picentini mountains, is made up of dense areas, such as Pompeii and Scafati, and of fruitful countryside even if differently urbanised. Its regeneration is based on the idea of green discontinuous city, including rural areas, thinking the Giuseppe Samonà's "extended city" over, related to the current urban problems. The proposal is a multipolar system, built up by small agro-urban clusters separated by rural areas and connected to the reclamation of Sarno river. The main goals of this strategy are to reduce sprawl, to improve soil and canal utilisation and to face overflowing risks and exodus needs.

Keywords Sarno river plain • Extended city • Multipolar system • Agro-urban cluster

9.1 The Sarno Plain

The analysis of the settlements along the Sarno river has been the occasion for a new green city plan extending in the Vesuvian countryside, including cultivated areas, natural landscape and small urban centres. Since the mid-twentieth century, the rural character of the agricultural plains in Campania has been affected by unplanned building in the countryside, transforming these areas into densely built-up suburbs or urban sprawl. To create alternatives to these two forms of development, both of which pose unresolved problems, this study aims to plan agro-urbanisation models providing sustainable solutions to the residential, productive and landscape need of peri-urban areas.

The Sarno valley has a long-standing agricultural tradition with a complex infrastructure bounded by a stunning backdrop of hills and mountains, with Vesuvius to the north and the Lattari and Picentini mountains to the south and to the east. The area has traditionally been used for specialist cultivation and manufacturing,

C.A. Manzo (✉)
Department of Architecture and Industrial Design, Second University of Naples,
Via S. Lorenzo, 31, 81031 Napoli, Italy
e-mail: carale.manzo@gmail.com

R. Ingaramo, A. Voghera (eds.), *Topics and Methods for Urban and Landscape Design*, Urban and Landscape Perspectives 19,
DOI 10.1007/978-3-319-51535-9_9

in particular the agricultural and food sector and nursery gardening. Currently, the plain has a dual nature which explains the uneven concentrations of buildings. On the one hand, it has succumbed to pressure from the urban area of Naples which has led to high-density areas on the coast and in nearby conurbations. On the other hand, it possesses the agricultural features of the Sarno valley with increasingly fewer buildings as one moves upriver further inland. Besides the coastal area around Castellammare di Stabia, other high-density zones of building include the urban sprawl between Pompeii and Scafati on the northern side towards Vesuvius (with the major attractions of the archaeological area and the Marian Shrine) and between Angri and Nocera in the foothills to the south. Besides these high-density zones, the Sarno plain currently resembles a series of urban centres set back from the course of the river (Striano, San Marzano, San Valentino, Poggiomarino) linked by built-up strips along the roads, separated by agricultural zones dotted with minute residential and non-residential buildings (houses, greenhouses, warehouses and factories) and criss-crossed by a dense network of canals. The Sarno valley therefore alternates between mainly urban and mainly rural areas. In the former area, to the west, intensive building has reduced the extent of the countryside, shut in between densely urbanised zones, while in the latter area to the northeast, the distance between built-up zones still leaves space and an opportunity for new landscapes that preserve an agricultural character (Renna 1980, Manzo 2012).

Research work has led to a reinterpretation of this settlement pattern. Although they have undergone significant changes over time and the architecture is of mainly low quality, it is believed that the areas along the Sarno river can be reassembled into a discontinuous green city that grows by acquiring pieces of countryside in an overall pattern that can be traced back to the concept of the “extended city” coined by Giuseppe Samonà. In a famous work written in 1976, Samonà argued that open areas and rural landscape had a special importance in terms of settlement and foundation potential for the construction of the new modern city. He underlined the need for regulations that would protect, in a strictly architectural sense, parts of the countryside, recognising the importance that rural communities and agricultural processes can contribute to creating a broader overall architectural notion of the city (Samonà 1976) (Fig. 9.1).

In terms of preserving the human, cultural and technical traditions of the local territory, an important role can be played by the visible historical traces within the Vesuvian plain (the roads of *centuriatio* and the old canals, the land divisions and the plots that hark back to the Roman subdivision of land and the layout of ancient Pompeii). These features have gradually led to the creation of the morphology of the Sarno area and can be considered in terms of their importance in projects designed to transform the territory.

The closely adjoining houses, greenhouses and fields have retained the traditional residential structure in these areas, surviving the disasters of unfettered urbanisation. This settlement pattern seems to have given inhabitants the capacity – and desire – to belong to both the urban and the agricultural world. The attraction of the city, which in provincial areas is regarded as a sign of progress and an improvement on rural poverty, seems to be compatible with the opportunities offered by

Fig. 9.1 Sarno river old route (1796)

agriculture, which is rightly considered an opportunity for development rather than a factor of marginalisation or backwardness.

9.2 A Discontinuous Urban System

The first aspect of the research is the reference to the concept of the city, which has become widely accepted following the Samonà contributions and the urban studies established by Rossi and Ajmonino. According to these authors, the urbanised countryside is imagined as a place that is a constituent part of the contemporary city, transferring the proposed concept of "città per parti" (like a patchwork city) to the territorial scale (Fig. 9.2). According to this hypothesis, the Sarno plain can be viewed as the basis of a discontinuous urban system spread over the countryside where existing settlements can be considered as districts, where green and agricultural areas represent productive parks and gardens and where public buildings can be identified in the historic structures and new infrastructure (archaeological areas, the Marian Shrine and modern Pompeii designed by Bartolo Longo, the Bourbon explosives factory, but also the new hypermarkets, the multiplex cinemas, etc.).

The debate about the future of the western city, unlike Third World megacities, is shifting towards a reduction in the scale of the large, densely built-up city. This is demonstrated by the de-growth of several American industrial cities which are trying to find the right degree of urban downsizing forced by the reduction in the number of inhabitants/workers.

Fig. 9.2 Millers on Sarno river in 1800

As a strategy for monitoring this downsizing, the residential zones are designed into smaller centres separated by green areas created by the demolition of existing architecture and the delocalisation of factories. The urban centres of southern Italy have never reached these levels of de-growth, but by anticipating similar processes, it is possible to imagine a discontinuous system that safeguards green areas and controls the development of small settlements that have not been incorporated into conurbations. The enhancement of the productive green zones can count on the growing potential of the *green economy*, in particular in the Pompeii area, which is a centre of excellence for nursery gardening in Italy. In order to envisage an *extended green city* in the Sarno plain, it is particularly important to pursue a strategy of discontinuity that monitors the growth and building density. The consolidated city grew by continuously adding new parts to old ones; nowadays, its development should take place by increasing the density of built-up areas, using the processes of stratification that have always existed in the history.

The second aspect concerns the role of natural elements and the landscape in the search for the agro-urban features for the new green city architecture. For a long time, the Sarno river used to be a structural element of settlements in the countryside (underlined by the pile dwelling of Longola near Poggiomarino until the nineteenth-century houses and factories along the river). Today, it seems to have no influence on land use and architecture. Indeed, it is considered a multiplier of pollution and an ecological disaster of the zone. With few exceptions, the modern era has witnessed few significant examples of architecture related to the river and the canals, either for the banks or for the waterfront. The Sarno has therefore been chosen for this study because it provides a privileged insight into the settlement problems of the plain and can be considered as an incubator of new relations between the river system and the surrounding areas. Considering the environmental cleanup programmes – which include making the river navigable once more – and nursery gardening, the redevelopment of the Sarno can play an important role in the process of restructuring the local area, reutilising the unit created for technical and formal reasons that was once part of the rural landscape.

9.3 Transformation Criteria

Besides the typo-morphological and historical layout (*centuriatio*) analysis, the research has also explored the main planning problems currently taking place in the main municipalities (in particular Pompeii and Scafati) and the activities carried out by the Sarno river basin authority which was set up in the 1990s (Fig. 9.3).

Instead of setting out to create an overall plan, the study has followed contemporary theoretical and methodological interests with the aim of identifying the criteria for transformation, working on several sample areas as general solution for the settlement problems in the plain. The work has emphasised the regulatory and "systemic" aspects rather than specific solutions with formalised projects, focusing on several issues that concern research in the whole of Italy: the role of smaller centres; the relationships between residential structures, free spaces and parks and

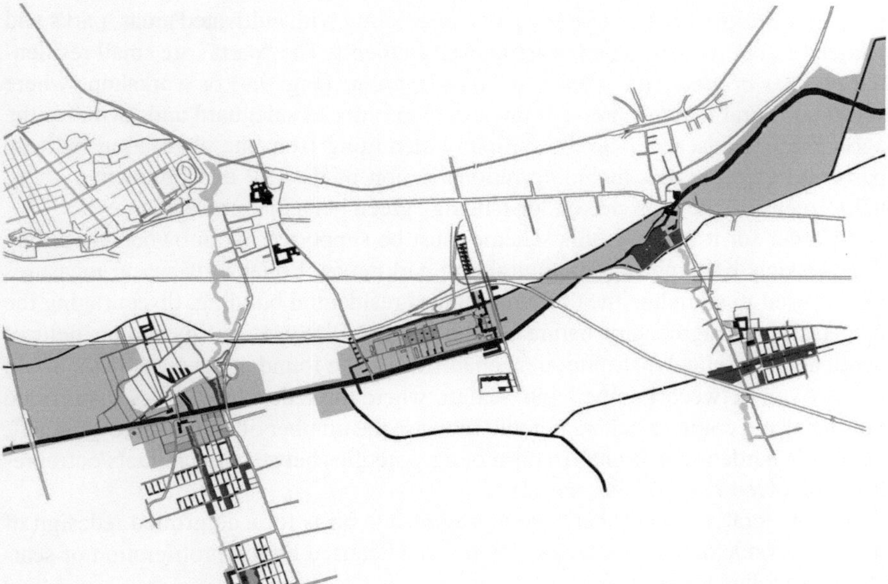

Fig. 9.3 Pendulum crossing the Sarno in Pompeii and Scafati

gardens; then the nature of workplaces; and the infrastructural system above the municipality scale.

The emphasis on theoretical and regulatory aspects over immediately achievable objectives also stems from the unequal town planning instruments of municipalities, the uncertainties of urban and environmental redevelopment initiatives and the discordant timescales of legislation for the safeguarding of the local area. Given the impossibility of implementing them in a short timescale, the proposed initiatives assume – and at the same time promote – the improvement in environmental conditions; in particular, they concern the redevelopment of the Sarno river for which, in its final stretch from San Marzano to the mouth, the results are still hard to discern. There are various restrictions, but it is worth highlighting the regional and provincial measures designed to prevent and limit risks from seismic and volcanic events which have recently extended the safety "red zone" to the entire municipalities of the towns concerned. According to these requirements, the initiatives are subordinate to two conditions: a higher level of structural performance to cope with the load of fine ash fallout from an eruption and an improvement in the road system to ensure a quick population escape.

9.4 An Anti-dispersal Strategy

As an anti-dispersal strategy, a discontinuous system of "low-rise high-density" initiatives, with mixed functions, has been adopted. The core idea is a green city made up of architecturally defined basic parts, alternating with cultivated areas, parks and gardens and of recognisable infrastructural elements. The "parts" are small residential *clusters of productive houses with greenhouses* (Fig. 9.4) or workshops where urban and rural features are well integrated in order to safeguard and revitalise the agricultural landscape. The discontinuity stemming from the alternating built-up parts and green zones, including rationalisation of the PIP and productive areas, helps to ensure the presence of "distancing" green belts that prevent sprawl.

In order for it to work, this scheme must be supported by an ownership policy that prevents building in agricultural and park areas. Equal distribution measures can be used to stimulate the concentration of residential building, discouraging the dispersal and fragmentary nature of individual initiatives. An interesting field of application for equal distribution mechanisms can be found in the areas to the south of the river, between Pompeii and Scafati, where the redevelopment of dismantled industrial and commercial areas could increase the number of "productive gardens", vegetable gardens and orchards, promoting synergies between inhabitants/entrepreneurs involved in gardening nursery.

The integrated cluster therefore represents the basis for a controlled redesign of the countryside aimed at slowing down sprawl caused by the proliferation of scattered individual buildings.

Given that the growth of these municipalities, which have been classified as the red zone in the security plan, is supposed to be "zero balance", the increases in

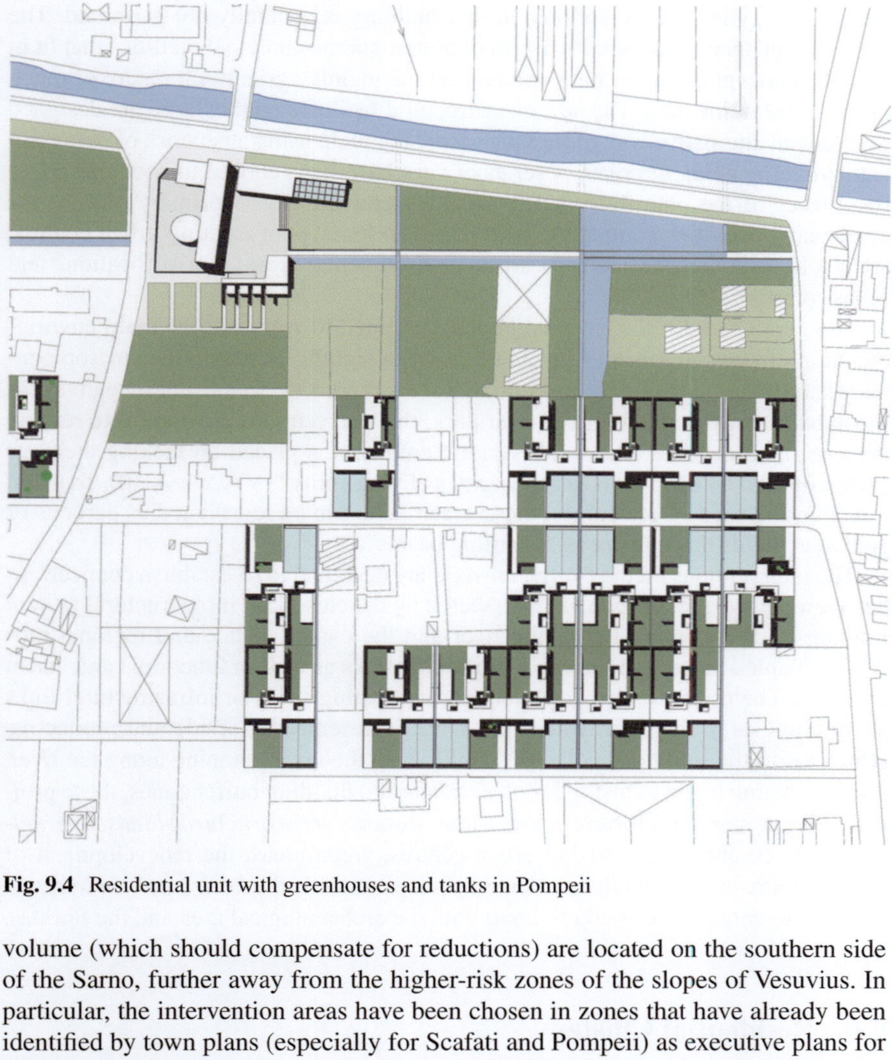

Fig. 9.4 Residential unit with greenhouses and tanks in Pompeii

volume (which should compensate for reductions) are located on the southern side of the Sarno, further away from the higher-risk zones of the slopes of Vesuvius. In particular, the intervention areas have been chosen in zones that have already been identified by town plans (especially for Scafati and Pompeii) as executive plans for redevelopment and/or amnesties for illegal building or in areas for productive settlements.

9.5 The Transverse Strategy

The intervention proposals follow two directions, a longitudinal axis along the river and another prescribing a transverse layout. The system of projects situated on a *longitudinal axis* along the course of the Sarno river takes account of flood risk areas where, with few exceptions, further building is generally not permitted. The aim is to enhance the river with a system of new supra-municipal facilities that fit in with the concept of a river park and respect the mainly agricultural nature of much of the surrounding area. The new infrastructural facilities on the Sarno are designed to create an alternative way of moving across the plain using a network of roads and paths running along the river. A series of sites have been chosen to become collective places of the plain intended to be a "patchwork city", "primary places" for condensing activities compatible with the main features of existing urban centres: archaeological sites, agricultural areas, environmental areas, tourist locations and fitness places.

The guidelines for the riverbed banks are aimed to work out the main environmental problems: in the built-up areas, the most significant regard the environmental rehabilitation or thinning-out initiatives, supported by equalising incentives. In the unbuilt areas planned for the river park, the interventions are careful to respect the rural landscape and the water system; emphasis is placed on making the river park proposed by the river basin authority a "living entity" (VV.AA. 2010). To avoid remaining unused safety valves, these green belts can be redesigned as gardens or cultivated fields that can act as "retarding basins".

The interventions on the *transverse axes* are designed to re-establish connections between the two sides of the plain dissected by the close-knit infrastructural system running in direction west-east (the river and the canals – state and regional railways – Naples-Salerno motorway), improving links as well as functional and visual relations. The transverse strategy consists in creating strips or infrastructural links at right angles to the river, whose poles are represented by residential-productive structures to the south and collective facilities to the north, running along the river. By reassigning a role to historic routes, canals and flooding buffer zones, these projects offer a new equilibrium for the area through *infrastructural-landscape networks* which challenge existing urban centres, green areas, the redevelopment of infrastructure and productive settlements (an example is the *horizontal cableway* in Pompeii to connect the southern areas with the archaeological area and the Shrine).

9.6 Residential Cluster

The theme of the densely built-up *residential cluster* is added onto a transverse "pendulum link" and reinterprets the clusters of traditional building types of rural Campania. The new houses offer greater functionality and flexibility in the relationship between buildings, greenhouses and open spaces. They provide a sustainable alternative to the usual solutions of isolated blocks of flats and terraced houses. Designed as single plans, the clusters are intended to promote cooperation by rationalising the production and distribution of fruit, vegetables and flowers. Based on the joint concept of a house which is both rural and urban, the study has explored the theme of the residential unit with various solutions concerning the relationship between the land, greenhouses and agriculture (Manzo 2012).

Fig. 9.5 Productive residential cluster in Scafati

1. An initial model of the residential cluster (Fig. 9.5) is based on the concept of the individual-productive house where workspaces (greenhouses or workshops) are arranged within a single plot and closely combined with domestic spaces. The concept of the small self-sufficient unit based on the autonomy (which may also be productive) of individual residences is designed to meet the widespread need of the single family or independent "parental" house. Each residential cluster consists of a house nuclei suited to the rural-urban nature of these areas, and the various-open plot character facilitates the inclusion of the pre-existing buildings

into the new residential structures. From the functional viewpoint, flexible modules make it possible to expand the basic residence by adding other residential units of various sizes.

2. In a second model, tried out in the southern area of Pompeii-Moregine, the greenhouses are centralised: the proposal is based on residential units made up of plots with gardens, arranged around a complex of greenhouses and fields running as a cooperative by the inhabitants. The composition principles underlying these green districts, protected from the river by *retarding basins*, feature large blocks marking the boundaries of productive complexes which are parts of formal-functional clusters within the layout of the extended city.
3. A particular model, the *water-based cluster*, has been tried out in the southern area of Scafati in areas of illegal building which need to be redeveloped or reconstructed: it tackles the problem of buildings raising from the ground, comparing residential schemes that can cope with the flooding risk. These morphological experiences have adopted the dwelling on *pilotis* as a solution for the house relationship with the land and water.

In contexts further away from the river, the theme of the dense cluster aimed to the green city has been explored by rethinking land use; the model involves housing, spas and structures for leisure activities designed to be closely linked to the natural environment in a mutually beneficial relationship.

The idea of the extended city, made up of various parts, centres and infrastructural lines, is not intended to be a global unitary design. New productive areas play an important part and are considered to be the "cornerstones" of the new green city.

References

Manzo CA (ed) (2012) Recinti metropolitani. La costruzione della pianura campana. Edizioni Kappa, Rome

Renna A (1980) L'illusione e i cristalli: immagini di architettura per una terra di provincia. Clear, Rome

Samonà G (1976) La città in estensione. In: Conference at the Faculty of Architecture in Palermo

VV.AA (2010) Basin authority of Sarno river report. Naples

Chapter 10
Work *in* Process, Work *on* Process: Designing the Riversides

Transformation and Participation Throughout the Architectural Project

Antonio De Rossi, Alessandro Armando, and Mattia Giusiano

Abstract "Linear" geographical objects such as rivers and infrastructure are increasingly proving to be interesting places to experiment with trans-scalar relational planning and design covering vast areas, which strive, not without difficulty, to go beyond the physical and nonphysical limits of each disciplinary field. This article examines such experimentation through two case studies in Italy: the *Progetto strategico della rete ecologica del Parco Fluviale Gesso Stura* ("Gesso Stura Riverside Park eco-network strategic project") between the Alps and the Po Valley in the province of Cuneo and the creation of the *Parco del torrente Sangone* ("Sangone River Park") near the metropolitan area of Turin.

Keywords Urban visioning • Urban strategic planning and design • Trans-scalar design

10.1 Introduction

It has been known for some years now that linear objects, whether geographical or man-made (roads, railways, rivers, etc.), are of particular interest, both from an analytical and design point of view. Crossing areas with very different physical features and morphogenetic rules, linear objects have the unique quality of making their internal logic and operating mechanisms comprehensible. When exploring regions by means of roads or rivers, the initial feeling of confusion and disorder, which arises from the transformations of the past few decades, is replaced by our awareness of conflicting rationale and independent and irreconcilable strategies. Each of them is limited and absolute at the same time: limited because it is based on

A. De Rossi (✉) • A. Armando • M. Giusiano
Department of Architecture and Design, Politecnico di Torino,
Viale Mattioli 39, 10125 Turin, Italy
e-mail: antonio.derossi@polito.it; alessandro.armando@polito.it; mattia.giusiano@polito.it

R. Ingaramo, A. Voghera (eds.), *Topics and Methods for Urban and Landscape Design*, Urban and Landscape Perspectives 19,
DOI 10.1007/978-3-319-51535-9_10

simplified and reductive concepts and absolute because it aims to have sole control over a plot of land by setting boundaries. Linear geographical objects, precisely because of their "narrative" ability to clearly show the reasons behind land-use changes, have an extraordinary potential in terms of design, as they offer an overview that goes beyond the individual rationale and strategies for each individual "regional fence".

Linear geographical objects, thanks to their analytical and design value, become an extremely valuable conceptual and operational "space", especially during a structural crisis of the welfare model and a profound redefinition of economic and design models behind the land-use transformation processes (Harvey 1990). It is precisely this crisis and this mutation that forces us to radically reconsider how land-use projects are carried out: the redefinition of the role of the vision, as a teleological device to orient long-term processes; the development of diachronic instruments to represent the transformation in its mid-term steps; the extension of the representational function of drawings and documents to other dimensions, beyond the mere morphological aspects; and the integration of the bureaucratic and technocratic aspects of the processes within the narratives and the figures of the architectural work (Amin and Thrift 2001).

10.2 Project Strategies and Reconsideration of the Project's Role

The idea of the morphological project that we may imagine is one that accepts the morphological and political conditions of the situation that we have so far described and tries to "internalise" them, transforming them into a positive factor and an active agent for project development. Working according to this perspective, however, implies compromising the traditional epistemological status of the project, which creates unease in the field of architecture (Olmo 2010). This means minimising that typical feature of modern projects: formal prophecies primarily pursuing their own self-fulfilment, the focus and legitimacy of which is based on this aspect alone. It means increasing the strategic and discursive value of the project, this being the instrument and conceptual space needed first of all to formulate and define objectives, scales, sizes and relational modes. It means focusing on *interpretation* rather than on *constructing*. Traces prevailing over taxonomy. This does not mean denying the assertive dimension of the project, its regulatory value. On the contrary, rather changing its character, reducing its self-referential and self-validating nature (Gabetti 1983).

At the same time, adopting this method when carrying out a project entails focusing on hidden and latent meanings, on possibilities to be made real. It entails displaying, *opening up*, by means of the indeterminateness of metaphorical images and their pre-political significance, following a line of research already outlined by geographer Giuseppe Dematteis, with his "poetic geography" idea (Dematteis 1985). This unveiling process also produces genuinely concrete results. Many of the criticalities in the design of contemporary landscapes are determined by the reductionist and field-focused way through which *problems are created*, even more than by not resolving functional and environmental issues. The issue of deconstructing apparent problems and rewriting questions – which refers once again to the centrality of the *jeu d'échelles* – can also have economic repercussions on the project.

This type of project design conceptualises the overcoming of incrementalism and its unlimited costs (economic, functional, environmental and political), rather than deferring the problem to an external political rationale, which can, in turn – in most architects' daydreams – legitimise an intrinsic rationale of form to which all parties must adapt. In an attempt to oppose it, if anything, it accepts and internalises its logic, its pulviscular nature and open time frame, in order to return it to a superior project process within a different framework.

This way of observing, therefore, sees the project as a kind of "negotiating table" or rather as an "outline" through which to develop design proposals for transformations. This pre-project phase – through the creation of a common language, that is also morphological – allows for the different positions to "merge" and, therefore, to become expressible, changeable and, ultimately, to converge. In this framework, the morphological project is not merely the end but also the *means* for carrying out transformations.

This type of project sees morphological landscape construction as the laying out of skeletons, reinforcing frames, patterns and structuring. To do this, it makes use of individual materials found in and on the land, with the aim not to impose additional forms on the space but to *bring out meaningful forms* in accordance with the existing processes and data (Secchi 1989, 2001). We need to understand and define which parts can be regarded as lasting structures and which are likely to change or, rather, can be restored to ecological re-naturalisation: for example, networks and lines are long lasting (they are the "strong" elements of the framework), while the "fields" of these imaginary frames may be a "softer" part of the pattern, which can change and transform themselves. This change in strategy, at a local policy level, has strong repercussions especially on transformation mechanisms, on the relationship between public entities and private stakeholders, as well as on project building and implementation methods, on the perception of the landscape and on the regulatory apparatus itself, which currently requires a radical revision (Lanzani 2003).

Understanding the value of very large-scale "environmental architecture" (water networks, ecological corridors, geomorphology, etc.) is, therefore, of fundamental importance. These are primary elements of long-lasting landscape design and are potential catalysts for recycling projects, able to restore the value of local environmental systems (Corboz 1983; De Rossi and Robiglio 1999; De Rossi et al. 1999; Durbiano and Robiglio 2003).

Recognition, *reconstruction* and *representation*, founded on the *centrality of morphology*, are pivotal to this project design method. The beauty and legitimacy of this process, however, does not derive merely from "structural" values. Its value is rather measured by its ability to stretch things, to foster awareness and, therefore, action. This is not a fixed approach but *focuses on* that ontologically focuses attention not on *objects* but on the structure of their *relationships*.

It is above all essential, however, to attempt to shift from a project concept primarily aimed at prescribing the quantity and quality of transformations, which follow a non-negotiable rationale – vouched for by a "culture of experts", based on conformity criteria – towards an *argumentative approach*, able to consolidate residential areas long before designing the spatial and productive organisation according to regulated procedures. To do this, the project must be able to strongly impact the technical and political *vision*, both in the sense of denaturalising the established ways of thinking about the issues – the ready-made and off-the-shelf solutions, rather than defining solutions starting with the problems – and with regard to using scraps of meaning as a starting point for the project. From this point of view, when viewed mainly as an instrument of expertise, the project has the primary responsibility – especially in a society like ours where phenomena are redundantly and precisely analysed but never short-circuited against political time frames and facts – to attempt to rewrite reality apparently seen as a given, starting from critically resetting project issues.

The following examples give an account of two attempts in academic design research to investigate the potentials created by using design instruments according to a negotiating approach and with changeable scenarios.

10.3 Work in Process: The "Gesso Stura" Riverside Park

Reasoning – and designing – in a *relational* way in very large-scale areas, where usually work is carried out by each disciplinary field separately, can involve considerable difficulties and must include an experimental phase.

In 2012, the *Ente Parco Fluviale Gesso Stura* (Gesso Stura Riverside Park Organization) commissioned a *Progetto Strategico di Rete Ecologica* (eco-network strategic project) from the Department of Architecture and Design (DAD) at the Polytechnic of Turin under the Alcotra 2007–2013-P.I.T. Programme "Cross-border Marittime Mercantour Region".

The project is especially interesting due to the very nature of the Park and to the part that the ecological network can play within the park. The PFGS (Gesso Stura Riverside Park) is a "work-in-process" park, resulting from a bottom-up aggregation approach, currently ongoing. This is one of Italy's few examples of a park created solely thanks to local administration (2006). It will only later be officially sanctioned by the regional body (2012) and is still open to new subscribers.

The management of the Park is also "atypical": no specific new institutional body was created, and it is managed by the Cuneo City Council that was behind the creation of the project. This is a streamlined structure that, however, has a limited range of operation – officially it cannot operate independently in areas belonging to other municipalities – and has limited spending powers. This is just one of the city's environmental department branches: few staff and little space. Therefore, since its creation, the PFGS has been forced to fund its activities through regional and European tenders, often in collaboration with the larger and "traditional"

Alpine parks: Maritime Alps, Marguareis and, in France, the Mercantour. This is the case with the Alcotra project and the "eco-network strategic project" funding. This collaboration immediately reveals the special nature of the PFGS area as compared to traditional parks: this is not a large range characterised by areas of environmental value with very little human presence but rather a linear area created by merging residual areas that are still undeveloped within a densely populated plain.

Evolving boundaries, limited operational potential and a densely populated area have pushed the Park to undertake a somewhat unorthodox path when putting into practice this ecological network: its design could be the first step towards defining an "appropriate" size for the Park and its potential range in the long run, going beyond the contingent conditions of the moment.

For this reason, among others, the Park Organization decided not to involve a team specifically focused on the environment but a pool of researchers within DAD focused on landscape design.

Over the years, DAD has studied transformations of settlements, infrastructure and environments in very large-scale areas (Secchi 2010; DIPRADI 2010a, b, 2011). Furthermore, DAD has taken part in a number of projects, which include the construction works for the Piedmont Region Landscape Plan (Regione Piemonte 2009).

In addition to that, DAD's working group had gained in-depth knowledge and respect in Cuneo province – where PFGS is located – by creating, a few years before, a *landscape vision* known as *(g)rand(a)stad* (De Rossi 2009; Giusiano 2012). It is a physical image of the area that, for the first time, highlights and correlates potentially key topics of landscape transformation – the "large-scale architectures" – able to create awareness and give a unified identity to a historically fragmented and polycentric area.

Three projects that cover different aspects: a local rail mobility system that uses an old disused railway network, known as *MetroGranda*, a green ridge around the Stura river (the most important central river in the province of Cuneo), an *agricultural heart* where agro-technical production and slow tourism can connect.

The "Green Ridge", the precursor of a possible expansion of the riverside park, up to then found only in the area of the city of Cuneo, became the point of contact between DAD and PFGS, providing a starting point for the expansion of the protected area to cover the entire fluvial channel. DAD and PFGS jointly decided to transform the ecological network project into a sort of Park Development Programme, going beyond field-specific issues and clarifying the main local issues.

The humanised environment, its linear shape, the fragmentation of environmental information and administration tools, indeed, led the research team to question the very nature of the ecological network concept in this situation. The most traditional reference paradigms for ecological networks seemed to involve a series of issues not always easy to resolve.

A *species-specific* network – aimed at specific target species according to their needs and functions, such as the interconnected system of habitats for biodiversity – was affected by the lack of uniform information available for the relevant area and by the difficulties encountered in integrating with the settlement and landscape transformation processes of the area.

An *eco-structural* network – zoning according to existing and mappable natural areas, or related to vegetation categories, defined also on the basis of conditioning macro-factors (geological substrate, local climate) – or operationally, a network of protected areas, as a park and reserve system and more generally governed, within a coordinated system of infrastructure and services, clashed with the lack of a single stakeholder able to coordinate and manage the entire area in question.

The research team has, therefore, decided to give the ecological network a *versatile* meaning: a "*medium term, multifunctional landscape ecosystem, defined on the basis of the previous functions and, more generally, in relation to human activities present in the areas in question*" (Malcevschi 2010; Malcevschi and Lazzarini 2011) (Fig. 10.1).

It is a way of conceiving the network as the creation of a green infrastructure that acts, at the same time, as a support for the habitats of different species and for human activities linked to its use and appreciation, both in terms of leisure and for the local economy. With this in mind, the network project is an instrument for dialogue – and not conflict – with the other planning instruments that determine the physical construction of the landscape: the network project presents itself as a potential strategic project for the region (Indovina 2009).

Dialogue is communication, and for communication to be effective – even outside the typically strictly regulated environment of ecological networks – it must be intelligible for most people. In parallel with the development of the network itself, its reconceptualisation is implemented by defining concise concepts able to enhance both the landforms and the related underlying strategies.

If observed on the large scale, the riverside landscape appears to be made up of three different geographical areas. Upstream and downstream areas cover a large range, where, in addition to the major fluvial channels, are networks of confluent minor streams and canals. The intermediate area, on the other hand, has a linear axis that, partly because of the sunken riverbed, is less connected with the surrounding areas.

All of these areas together create a riverside *tree-like* landscape, where the upstream area of the Gesso-Stura confluence forms the *roots*, the linear stretch between Cuneo and Fossano the *trunk* and the area between Fossano and Cherasco the *foliage*.

This image enables us not only to immediately understand the area's shape but also to briefly define possible large-scale strategies, while in the two areas appearing as the *roots* and *foliage*, it is fundamental, first of all, to strengthen the cross-connections between the different fluvial channels, in the central stretch protection interventions will mainly focus on broadening the primary corridor along the Stura.

In addition to these "local strategies" (because they are based on the specific characteristics of the different situations) the plan pursues a set of cross-cutting "general" strategies organised around three axes: strengthening the green infrastructure, redeveloping and defragmenting the existing habitats and a sustainable use and

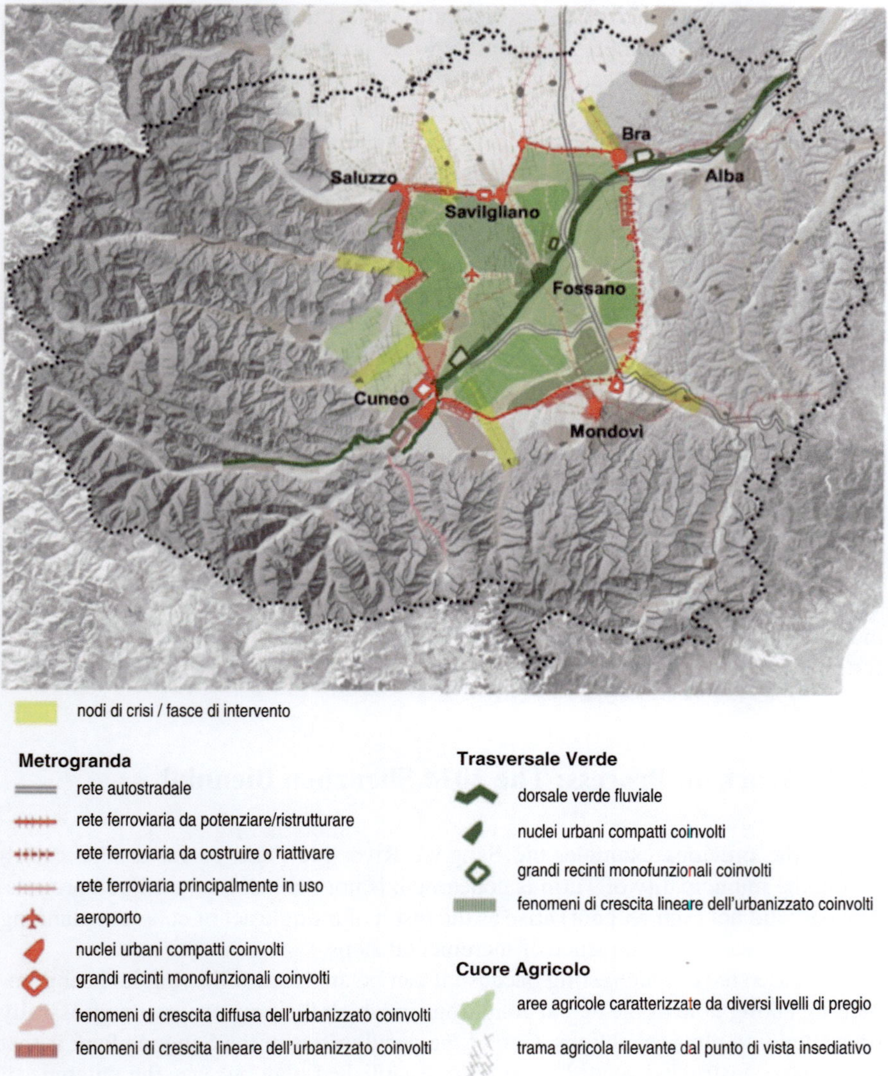

Fig. 10.1 A new vision for the province of Cuneo: *(g)rand(a)stad* (Source: Antonio De Rossi 2009)

enhancement of the natural and cultural heritage. These two groups of strategies together thus allow us to precisely define specific guidelines and actions.

An image of the landscape, three landscape contexts, three cross-cutting objectives: the strategic project for the ecological network seeks to create a common vision for a large-scale area able to identify – and adapt to its needs – a mosaic of specific requirements and conflicts, whether "locally specific" or "field specific".

Interpreting the "landscape" as an outline and the "landscape project" as the creation of the stakeholders that transform the landscape, the ecological network project pushes its field-specific nature to focus on wider issues and needs.

It does so for two reasons. First of all, because the project (also) involves areas over which the Park does not (yet) have jurisdiction: it either manages to convince the stakeholders outside the park or it is simply ineffectual (Fig. 10.2).

Secondly, because it aims to immediately defuse any conflicts related to environmental protection in densely populated areas, highlighting the points of contact between the needs of the environment and those of the manufacturing sector.

However, this approach is not always understood. Many environmental technicians see dealing with local needs as bending to external interests, thus misinterpreting an instrument that in some ways has a strong European status. The clients themselves, after an initial fascination, have trouble pursuing such an approach because of its real-life application: who wants to carry out concrete actions when they are not restricted to one single disciplinary field (or budget)? Who makes the first move to pre-order subsequent steps, taking into account that the will of individual local stakeholders can change over time – even for internal reasons – thus impacting the project itself?

Again, it is clear what probably is, and has been, the main issue for very large-scale projects in Italy in the past few decades: despite a general improvement in local *governance*, these projects have suffered from a congenital lack of support for effectively implementing projects, that otherwise too often remain on paper.

10.4 Work *on* Process: The 2014 Shenzhen Biennial

Unlike the previous example, the Sangone Riverside Park, as far as the stretch within the municipality of Turin is concerned, is not the result of a unified process. The park did not (and will not) arise as the result of a single act of coherent planning but out of a very long sequence of incremental steps.

The case study is interesting because it can be analysed following its near completion, taking note of the actual conditions in which the projects were successfully translated into physical actions. One of the peculiarities of the Sangone Park is that it was not constructed according to a pre-established plan but was the outcome of many predominantly technical and bureaucratic actions. None of which detracts from the importance of architectural design, which is linked to the effectiveness of the actions carried out but changes, as it were, their strategic function.

In 2014, the park project was the subject of an exhibition at the Shenzhen Biennial, which compared various examples of the transformation of urban river areas. Process outlines emerged, enabling comparison between different scenarios, often quite dissimilar. Such outlines are not analytical tools but (non-neutral) informative programme summaries and are almost exclusively retrospective as they describe processes carried out, rather than projects to be undertaken. Nevertheless, the discussions and exchanges between research groups following the exhibition

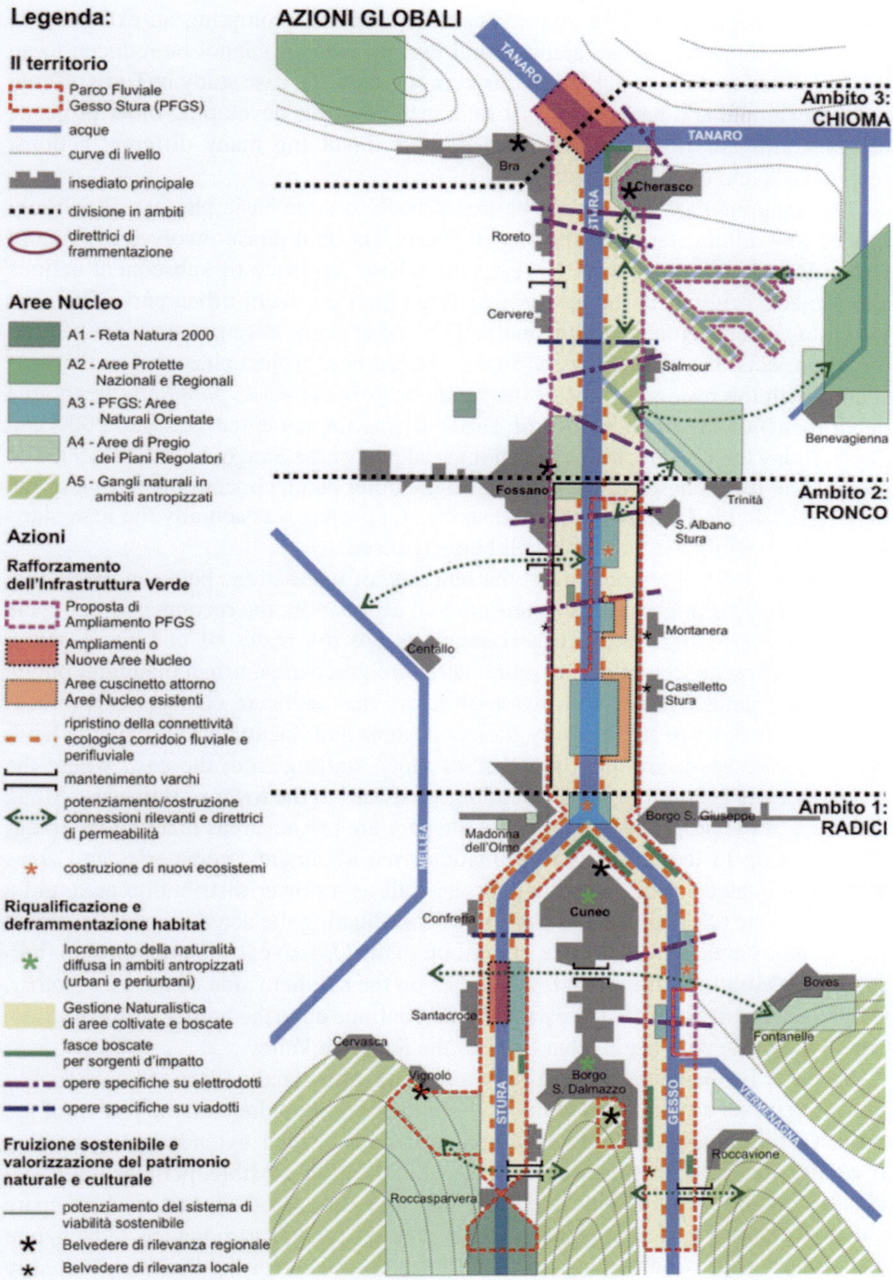

Fig. 10.2 The tree-like structure of the network project (Source: DAD 2011)

provided an opportunity for some general reflections, prompting an examination here of the impact of urban architectural design, when it cannot be reduced to an action limited to a single act. From this point of view, the case study in Turin offered a good example and has served as a reference point for developing other projects, demonstrating the logic of a lengthy process, involving many different actions, reappraisals and conflicting actions.

The Sangone Park project can be traced back to three basic phases, all publicly funded and administered by the city of Turin. The first phase involved the Artom Urban Renewal Project (URP), where the whole sequence of subsequent actions and projects originated, focused primarily on the Colonnetti urban park. The park was established before the Sangone project. After some disputes that blocked the initial projects that were proposed from above, a new project phase began that was shared with the residents, mostly run by the *settore del verde pubblico* (green area department) of the municipality of Turin. It was implemented between 2004 and 2006. Today the Colonnetti Park is an integral part of the Sangone Park Project. The second and third phases of the project, on the other hand, concern the design of the park itself, divided into two lots. The second lot project was actually the first, starting in 2002, while the first lot is still being realised.

From the start, the process was the outcome of interactions between green area projects, the liberation of areas, their gradual acquisition, the reconnection of cycle paths, river crossings, etc. These operations are the result of combined action between different sectors of city administration: green area, urban planning, public land/infrastructure, along with external actors that replicate changes and relocations, on the basis of fragmentary project designs and documents. Today, the land-use plan foresees a grouping together of plots, starting from the east, where the second lot has been set up, and extending westward to the town of Beinasco. Along the route of the proposed expansion of the park are private areas that will be gradually given up to the city of Turin, unauthorised allotments, cemeteries and areas already allocated to the riverside park, as well as a power distribution area and a cemetery. The basic aim of the city that is coordinating the acquisitions is to create a continuous area along the river, at least up to the Drosso castle park, but also linking to the Mirafiori Sud district, which lies on the southern side of the Fiat factory. After the Drosso green area, the park should continue up to the border with Beinasco, where the cycle path begins that leads to the Sangone Valley.

To describe the complexity of such a process, attempted with the Shenzhen diagrams, first met a spatial and temporal boundary problem. Indeed, the phases of the project presupposed multiple other preconditions, which extended into the surrounding area, and that made its contingent realisation possible, perhaps not traceable to any all-encompassing scenario. The four cases studied, however, compare and contrast very different places, in Italy and in China (Turin, Venice, Guangzhou and Hong Kong). Considering the other cases described for the Biennial, it clearly emerges that complex projects, even under highly authoritative regimes – as cities in China could be considered to be – are unavoidably subject to continual changes of direction, alterations and negotiations, with repeated redesigns (Armando et al. 2015).

Fig. 10.3 Sangone Park Project (Source: Author's elaboration)

Projects of this type can thus be described in terms of results, rather than intentions, and their coherence is attributable more to the official and bureaucratic aspects that successfully call for concrete action, rather than grand narrations of scenarios. In other words, it is possible that transformative actions, including effective ones, are carried out using the project differently than through the creation of a unified *vision*, and above all that the impact of the projects may also be assessed positively, even when the result differs profoundly from initial projections (Albrecht and Balducci 2013) (Fig. 10.3).

What then is the function of the scenarios and project narratives, assuming that they can be systematically overruled? The hypothesis emerged in all four cases that the scenarios were made to be altered, and the alterations to projects facilitated their completion. Because it is through a series of proposals, objections, renegotiations and redesigns of the projects that the necessary interpersonal relations were achieved for collective action (Armando 2015).

Drawing up the process diagrams had the aim of examining the chain of responses that transform the project designs into concrete action, obviously, starting from a retrospective analysis but with the intention of extending into the future the possibility of using those diagrams. Indeed, this very possibility seemed the key to the entire job: the extrapolation into the future of a project's chain of effects, including its

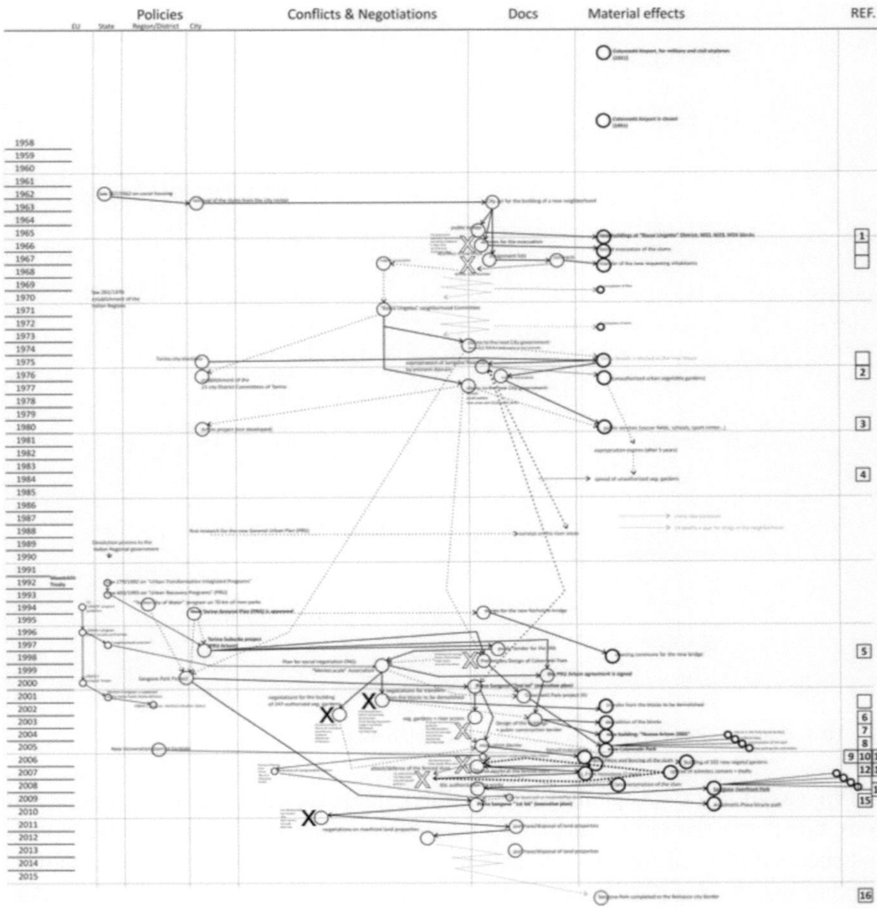

Fig. 10.4 Shenzhen diagram, Sangone Park (Source: Author's elaboration)

probable modifications, becomes a strategic aspect of project activity, which could be integrated into the development of the project itself. Of course, the fundamental problem of a chain of effects extrapolated into the future is the unpredictability of what lies ahead (Fig. 10.4).

Some conditions may be considered predictable, allowing a diagram of the future to be defined, such as when (1) a process has already started, with standard procedures, that allows us to identify (some of) the future steps required; (2) there are contract deadlines (schedules, permits, financing, etc.) that define the *termini ad quem* able to foresee decisions before they are taken, with multiple consequences; and (3) there are concurrent predictable factors (provisions for various sectors, constraints and regulations that are pending but not yet taken into account) that, when considered together, allow future and seemingly indeterminate scenarios to be inferred. But there are also other conditions that occur in an unpredictable way and which depend on intentional actions taken by those involved in the process (decisions, individual and collective choices, objections, protests, etc.). There are thus many points of divergence before a particular chain of effects can occur, if looking to the future, which depends on the inscrutable way in which individuals decide to act.

We can imagine that to plot the effects of a process into the future, along the lines of the descriptions tried out in the Shenzhen diagrams, basically means trying to pursue hypotheses that bring together predictable and unpredictable factors, along lines of possibility. However, the approaches linked to developing a scenario may be many, and they affect the way in which planners believe they can influence events through their own tools. It is possible to distinguish at least two possible groups of approaches that emerge when developing a strategic project: (a) a predominantly positive attitude, confident of containing the unpredictability of the project's circumstances, and where the aim is to defend the initial coherence of a proposal; and (b) a pragmatic and "flexible" attitude, in which it is accepted from the outset that designs may undergo even profound changes and revisions, provided that in the end a tangible result is achieved.

The first approach could be described as a *low predictability project*: ideally, if all factors were predictable, it would be sufficient to simply extend the design problem to its proper scale within the system, calculating the trajectories of all effects to achieve the objective. This attitude encourages the indefinite extension of descriptions, the collation of masses of information, within the idea that tends towards total aggregation (one thinks of the rhetoric surrounding "big data" for example) it corresponds to an increase in project feasibility. The more descriptions of phenomena we are able to put together and calculate, the more the expected final outcomes will approach those actually pursuable. In this case, the project can be considered as having *limited rationality* (for lack of time, resources, etc.), without compromising confidence in the ability to define the process through the determinations of the project itself. Certain positions on participatory and "inclusive" planning also feed, in the end, the same premise: that there is a framework, perhaps infinitely complex, comprising economic factors, procedures, functions, infrastructure, preferences and performances that, if reconstructed, would give a very similar result to that actually achievable.

The second approach can be defined as a *low unpredictability project*: the rationality of the project is not only limited but is also divergent, because it continually produces new choices and inventions. The conditioning factors involved in the modification of outcomes are generated in the project itself, during its processing and socialisation cycles. In these predicaments arises the set of choices that cannot in any way be anticipated (similar to what happens in elections: individual choice is unpredictable). It is evident that in this second case, the project is seen as a device that creates its outcomes during the process and not as a coherent act that is placed at the start of transactions. Even the forms of inclusion and participation change in relation to this view, as new conditions, new actors and new problems are defined, the project takes on a different form and embraces new possibilities. The changes that project configurations express are no longer a surrender in the face of reality but a flexibility of scenarios required to connect a growing number of instances, recombining them cyclically.

The project can, even, bring together determinable factors (because these also exist) and narratives about the future, presenting itself as a device that creates progressive agreements, increasingly rigid and stipulated. One of the fundamental problems concerning strategic plans is precisely that they are not able to define what the procedural and contractual conditions will be that will gradually be imposed in light of a possible implementation. The move from scenarios to procedures often discounts a separation of the figural development phase, associated with agreements concerning the symbolic and form aspects, from the phase of execution and bureaucratic checks. In this way, the symbolic and bureaucratic exchanges remain independent, condemning large-scale plans and projects as unfeasible. On the other hand, one could hypothesise that tracing the possible configurations of a region's future physical shape is essential for intersecting administrative and technical issues, with narratives that prioritise instances and values. In this sense, the architectural project could bring together figures and procedures many times, on different scales, progressively setting an increasing number of conditions, which can be realised incrementally. It is a question of imagining a trans-scalar and diachronic project, which presents itself as a flexible frame of reference. A tool that, by designing the shape of things, is able to repeatedly reabsorb many modifications through processes that, inevitably, occur in the long term and that are fragmented by the administration of land transformation.

References

Albrecht L, Balducci A (2013) Practicing strategic planning: in search of critical features to explain the strategic character of plans. DisP – The Planning Review 49(3):16–27

Amin A, Thrift N (2001) Cities. Reimagining the Urban. Italian edition: Amin A, Thrift N (2005) Città. Ripensare la dimensione urbana. Il Mulino, Bologna

Armando A (2015) The XXI century periphery: Turin and the urban crisis. In: Armando A, Marina O (eds) Projects for an inclusive city. Social integration through urban growth strategies. Europe for Citizens Programme 2007–2013, Skopje

Armando A, Bonino M, Frassoldati F (eds) (2015) Watersheds. A narrative of urban recycle. Sandu Publishing, Guangzhou

Corboz A (1983) Il territorio come palinsesto. Casabella 516:22–27

De Rossi A (ed) (2009) GrandeScala. Architettura, Politica, Forma. List, Barcelona

De Rossi A, Robliglio M (1999) Nuovi ruoli per il progetto. Esplorazioni progettuali nei luoghi della trasformazione recente. In: Portoghesi P, Scarano R (eds) Il progetto di architettura. Idee, scuole, tendenze all'alba del nuovo millennio. Newton Compton, Rome

De Rossi A, Durbiano G, Governa F, Reinerio L, Robiglio M (eds) (1999) Linee nel paesaggio. Esplorazioni nei paesaggi della dispersione. Utet, Turin

Dematteis G (1985) Le metafore della terra. Feltrinelli, Milan

DIPRADI (2010a) Indirizzi per la Qualità Paesaggistica degli Insediamenti Buone Pratiche per la Pianificazione Locale. Research group: Bazzanella L, De Rossi A, Castelnovi P, Berta M, Delpiano A, Dini R, Giusiano M, Rolfo D. L'Artistica Editrice, Savigliano

DIPRADI (2010b) Indirizzi per la Qualità Paesaggistica degli Insediamenti Buone Pratiche per la Progettazione Edilizia. Research group: Bazzanella L, De Rossi A, Castelnovi P, Berta M, Delpiano A, Dini R, Giusiano M, Rolfo D. L'Artistica Editrice, Savigliano

DIPRADI (2011) Progetto strategico per la realizzazione della rete ecologica fluviale e perifluviale nell'ambito del parco fluviale gesso e stura e territori limitrofi. Research group: De Rossi A, Governa F, Barbieri M, Delpiano A, Giusiano M, Toldo A. Gesso Stura Riverside Park Organisation, Cuneo

Durbiano G, Robiglio M (2003) Paesaggio e Architettura nell'Italia Contemporanea. Donzelli, Rome

Gabetti R (1983) Progettazione architettonica e ricerca tecnico-scientifica nella costruzione della città. In Gabetti R (1997) Imparare l'Architettura. Allemandi, Turin

Giusiano M (2012) CittàGranda. Forma, immaginario e progetto del territorio alla grande scala. PhD thesis, DIPRADI, Turin

Harvey D (1990) The condition of postmodernity. Blackwell, New York. Italian edition: Harvey D. La crisi della modernità, Il Saggiatore, Venice

Indovina F (2009) Dalla città diffusa all'arcipelago metropolitano. Franco Angeli, Milan

Lanzani A (2003) I paesaggi italiani. Meltemi, Rome

Malcevschi S (2010) Reti ecologiche polivalenti. Il Verde Editoriale, Milan

Malcevschi S, Lazzarini M (2011) Messa a punto di un modello operativo per la raccolta di buone pratiche per la Rete Ecologica Regionale in Lombardia. ERSAF e Regione Lombardia, Milan

Olmo C (2010) Architettura e Novecento. Diritti, conflitti, valori. Donzelli, Rome

Secchi B (1989) Un progetto per l'urbanistica. Einaudi, Turin

Secchi B (2001) La forma della Città. In: Diario di un Urbanista. Available at: www.planum.net. Accessed 29 June 2016

Secchi B (ed) (2010) On mobility. Marsilio, Venice

Chapter 11
Microclimate Models for a Sustainable and Liveable Urban Planning

Guglielmina Mutani and Francesco Fiermonte

Abstract Globally, the 54% of the world's population reside in urban areas, and in 2050 the projections are of 66%. Then the sustainability and livability of urban spaces are rising the attention of the scientific community. Particularly, in this work, the microclimate of outdoor spaces is investigated considering the different outdoor air temperatures registered by various weather stations in the city of Turin (Italy). The air temperature variations were correlated with the characteristics of the different spaces as the built urban morphology, the solar exposure of urban spaces and the albedo coefficients of outdoor surfaces. Finally, with a multiple linear regression analysis, the air temperatures have been correlated with the urban variables to obtain a model for the prediction of the average monthly temperature in the city of Turin. This model will be used to understand the different microclimates in Turin but also to evaluate the most influential urban variables on the air temperature. Moreover, the resulted model could help urban planners to predict the microclimate in new districts as a function of the urban form and of the outdoor materials chosen.

Keywords Microclimate • Urban heat island • Urban planning • Air temperature model

11.1 Introduction

Nowadays, people live more in urban areas (54% in 2014) than in rural areas, and this trend is still growing with a prevision of 66% in 2050. The most urbanized regions include Northern America and Europe (respectively, with 82 and 73% of

G. Mutani (✉)
Department of Energy, Politecnico di Torino,
Corso Duca degli Abruzzi, 24, 10129 Turin, Italy
e-mail: guglielmina.mutani@polito.it

F. Fiermonte
Interuniversity Department of Regional and Urban Studies and Planning,
Politecnico di Torino, Viale Mattioli, 39, 10125 Turin, Italy
e-mail: francesco.fiermonte@polito.it

R. Ingaramo, A. Voghera (eds.), *Topics and Methods for Urban and Landscape Design*, Urban and Landscape Perspectives 19,
DOI 10.1007/978-3-319-51535-9_11

people living in urban areas), and all regions are expected to urbanize further over the coming decades (United Nations 2014).

Then, the control of the microclimate of urban environment became a very important matter as it can contribute to a better livability of the cities with an improved use of the outdoor spaces. In big cities, this phenomenon is strongly linked to the urban heat island (UHI), and these aspects are treated in the first part of this work as state-of-art and literature review.

The second part of the work is dedicated to the evaluation of various factors influencing the air temperature and its variations in the urban context of Turin. Turin is one of the most populous cities in the northern part of Italy with an interesting various territory with parks, rivers, a historical centre, suburban areas and industrial zones.

The aim of this work is to elaborate a model for the prediction of air temperature as a function of urban variables, solar exposition and the characteristics of outdoor surfaces. Particular attention will be paid to factors influencing or closely related to the urban heat island, as already underlined in the previous works.

Different tools have been used to evaluate the microclimate at urban scale. ESRI ArcGIS 10.3 has been utilized to calculate the urban characteristics with a detailed representation of the built environment, starting from the Technical Map of the Metropolitan City of Turin. For the evaluation of the characteristics of outdoor urban surfaces, several images coming from the Aster satellite have been processed. Finally, five different zones in Turin with a weather station inside have been selected to elaborate a model to predict the air temperature as a function of the urban form, the solar exposure and the albedo coefficient of outdoor surfaces. The resulting model can be useful in the design phase in order to plan a more sustainable urban development or to control air temperature variation in the urban space. The model can also help in understanding how temperature changes depend on urban planning variables, as the presence of rivers and green surfaces, giving indications on the interventions that can mitigate the overheating effect.

Some of the results show the urban heat island effect with lower temperatures near the parks and the riversides and higher temperature in the high-density centre and industrial zones. These results will be further investigated on a larger zone around Turin considering more weather stations, other satellite images as the Landsat ones and additional indicators, for example, the normalized vegetation index (NDVI).

The analysis conducted in this work starts from previous researches on the influence of different materials used for outdoor urban spaces on the microclimate of urban environments (Mazzotta and Mutani 2015), on the evaluation of the urban form of buildings' heritage and its influence on urban microclimate and heating energy consumptions of residential buildings (Delmastro et al. 2015a, b) and on the urban heat island effect in Turin (Mutani 2016; Mutani et al. 2016).

11.2 State of the Art and Literature Review

The urban overheating problem is a matter of fundamental importance that all the developed and the developing countries are studying to mitigate its effects. For big cities, this issue is strongly linked to the thermal phenomenon called urban heat island (UHI).

Starting from the 1970s, thanks especially to the availability of data, tools and techniques were improved, and new algorithms and computational models have been formulated to understand and analyse more deeply the heat island phenomenon. The first studies on this phenomenon were based on the acquisition of satellite images (i.e. Landsat) to calculate the land surface temperatures (LSTs) and to observe how this parameter changes according to the different types of urban surfaces.

Several studies (Aniello et al. 1995; Iino and Hoyano 1996; Jusuf et al. 2007) analyse the relationship between land use and land cover (LULC) to understand how human activities can influence the intensity of the heat islands. The first results show that the heat island is mostly concentrated in urban areas, while the phenomenon is not observed in the surrounding rural areas (Li et al. 2009; Unger et al. 2010; Onischi et al. 2010): this is why the temperature measured in urban context is higher than in the peripheral zones.

Later studies begin to take into account different variables and factors that may affect the intensity of UHI as the relationship between land use and type of coverage (LULC), the normalized vegetation index (NDVI) and the land temperature surface (LST). All these factors were calculated by elaborating satellite images, such as Landsat and ASTER, through the use of specialized software capable in the localizing information, such as geographic information system (GIS) (Weng et al. 2004; Onischi et al. 2010; Li et al. 2011; Zhang et al. 2012, 2013; Xu et al. 2013; Effat and Hassan 2014; Kong et al. 2014; Shahidan et al. 2015; Fernàndez et al. 2015).

Together with these studies, a very important aspect to be taken into account is the variations of microclimate and UHI with the urban morphology. Depending on the different types of buildings and urban forms, the urban heat island has different intensities. Some researchers have investigated the changes of UHI with the urban characteristics, i.e. the urban density, the urban form and height of buildings (Aniello et al. 1995; Iino and Hoyano 1996; Weng et al. 2004; Unger et al. 2010; Li et al. 2011; Houet and Pigeon 2011; Zhang et al. 2012; Schwartz et al. 2012; Merbitz et al. 2012; Li et al. 2012; Chun B and Guldmann 2014; Ivajnsic et al. 2014; Allegrini et al. 2014; Kong et al. 2014; Sailor 2014; Dabaieh et al. 2015). These parameters are also used to describe the urban canyon phenomenon, characterized by high buildings facing on the same street. In this condition, the solar radiation is almost completely absorbed by urban surfaces, and the air stagnates, with an increase of air temperatures and, consequently, the urban overheating. As well as with these parameters, also the characteristics of materials, as the albedo coefficient, should be considered. This parameter is an intrinsic characteristic of the materials and represents the property of a material to reflect incident solar irradiation. In fact, a material such as asphalt, with very low of albedo value, tends to store and absorb all the incident solar irradiation, reaching higher surface temperatures than clay materials that absorb less heat and have less high surface temperatures.

Many studies evaluate algorithms to calculate albedo coefficients using satellite images on different electromagnetic bands (Li et al. 2009; Stathopoulou and Cartalis 2009; Mallick et al. 2013; Kolokotroni et al. 2014; Santamouris 2014a ; Salata et al.

2015; Thophilou and Serghiedes 2015; Qin 2015; Coseo and Larsen 2015, Pisello 2015).

Nowadays, to investigate the UHI effect, the microclimate and the outside air temperature variations, the urban morphology, the properties of the open space materials (with the albedo coefficient) and the surfaces' temperature have been analysed (Effat and Hassan 2014; Balogun et al. 2014; Zhang et al. 2014; Pisello and Cotana 2014, 2015; Santamouris 2014b; Fernàndez et al. 2015; Yang et al. 2015; Touchaei and Wang 2015; Debbage and Shepherd 2015; Kaloustian and Diab 2015; Fernàndez et al. 2015; Berger et al. 2015. Santamouris et al. 2015; Quan et al. 2015; Coseo and Larsen 2015; Mirzaei 2015; Raghavan et al. 2015; Peron et al. 2015; Yang et al. 2015).

While almost all of the analysed documents are proposing to investigate the heat islands and their effect on urban temperatures, other studies are focusing, instead, on the influence of the overheating on thermal comfort and on energy consumptions (Balogun et al. 2014; Zhang et al. 2014; Santamouris 2014a, b; Pisello 2015; Gracik et al. 2015; Santamouris et al. 2015; Misni 2015; Liu et al. 2015). These studies focus more on the heating than the cooling energy uses; obviously, the outside temperatures strongly influence the energy consumptions for space heating and cooling, but also the electrical consumptions are influenced by the outside microclimate. These researches evaluate solutions to reduce the heat island effect, air pollution, energy consumptions and costs. The main ideas concern the use of mitigation techniques to reduce the air temperature in urban areas. One method is to replace the traditional roofs with "cold roofs" made from tiles or slates in clay. In fact, these surfaces guarantee higher albedo coefficients able to reflect the solar irradiation with lower temperatures (Pisello and Cotana 2014, 2015; Dabaieh et al. 2015).

Other studies are focused, instead, on the use of the green on buildings' roofs or facades or on parking structures; through evapotranspiration, external temperature is lower (Zhang et al. 2014; Kong et al. 2014; Fernàndez et al. 2015; Qin 2015; Raghavan et al. 2015).

It was also analysed the so-called cold asphalt, the last generation of asphalts with very-high albedo values, able to reflect solar irradiation and keep lower temperatures than traditional bituminous asphalt (Carnielo and Zinzi 2014; Qin 2015).

In almost the cited studies, the relationships between the outside air temperatures and some of the parameters mentioned above are investigated. Some of them provided also a model or an algorithm, to be used in the design phase, in order to create a more sustainable urban planning (Iino and Hoyano 1996; Unger et al. 2010; Yeo et al. 2013; Chun and Guldmann 2014; Feng et al. 2014; Ivajnsic et al. 2014; Allegrini et al. 2014; Zhang et al. 2014; Kolokotroni et al. 2014; Fernàndez et al. 2015; Quan et al. 2015; Morris et al. 2015).

In this work all the parameters characterizing the urban morphology, the solar exposure and the albedo coefficients will be analysed in relation to air temperature variation to define an algorithm for the evaluation and prediction of the microclimate in Turin.

11.3 Case Study: Turin

Turin is the fourth most populous Italian town in the North-West part of Italy with a particular territory characterized by surrounding mountains and hills, by many rivers and parks and an elegant built context.

Turin is called "the city of four rivers" because it is localized in a plain surrounded by Stura di Lanzo, Sangone and Po rivers with also the Dora Riparia flowing close to its historical centre. Its territory is also surrounded by the Alpine mountains, also connecting the city with France. Moreover, Turin is the Italian city with more public green with about 21.1 m^2 of green per capita and about 160,000 trees along the streets and in the parks.

The climate is continental temperate with cold-dry winters and hot-humid summers, and, as the majority of urbanized territories, Turin is characterized by the UHI phenomenon with significantly higher temperatures than in the rural and hilly areas around the city. Considering the weather stations inside the town, there are differences on the monthly average air temperature principally due to the buildings' density, the presence of industrial zones and the proximity of parks and rivers.

From the 1990s, the summers have registered a significant warming, especially from the year 2000; the maximum air temperature was recorded in the hot summer 2003 with a peak of 39.7 °C in August 11.

In Fig. 11.1, the monthly average air temperatures registered by the ARPA weather stations in Turin from June 2005 to June 2015 are represented (ARPA is the Regional Agency for the Protection of the Environment of Piedmont Region).

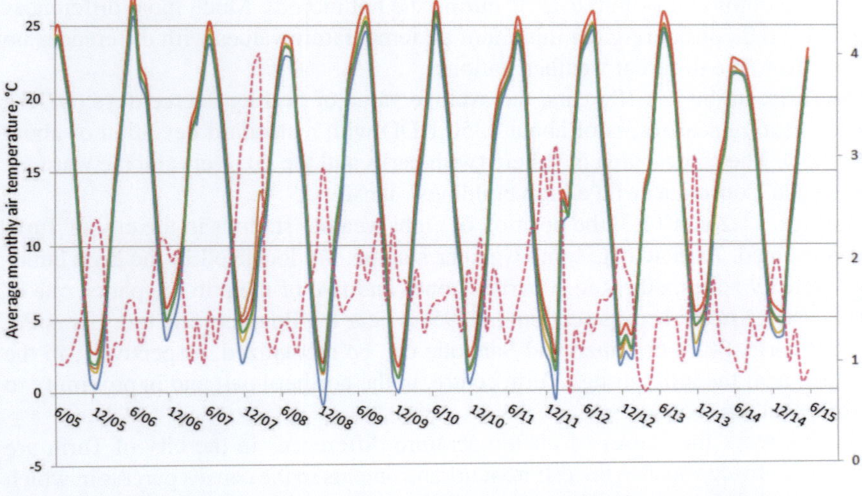

Fig. 11.1 Average monthly air temperatures registered in five ARPA weather stations in Turin

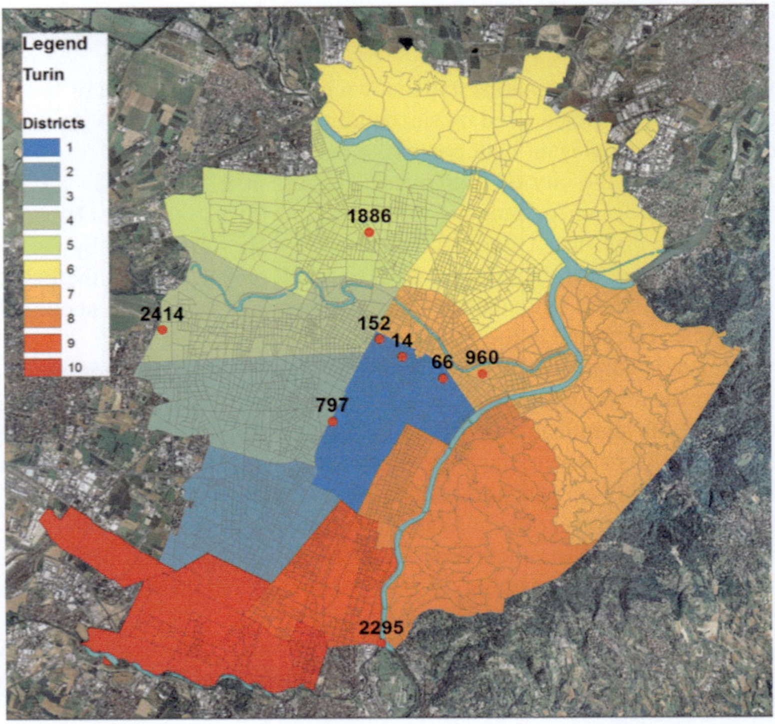

Fig. 11.2 The eight weather stations in the city of Turin with the indication of the relative census sections and the ten districts (Source: author's elaboration)

As it is possible to note, the temperature differences vary from 2 to 4 °C during the cold months and from 0 to 2 °C during the hot months. Much more differences are registered considering the minimum air temperature values with differences up to 9 °C from the different weather stations.

In Turin, in the last 10 years, the average value of heating degree days (*HDD*), for the heating seasons, is of about 2350 HDD with a standard deviation of about 200 HDD. The coldest area is Vallere (with parks and the Po river) and the warmest is via della Consolata (with a high buildings' density).

In Figs. 11.2 and 11.3, the position of eight weather stations in the city of Turin is represented. Particularly, some weather stations are localized in the high buildings' density zones, others in suburban zones and two of them in the parks; one of these, Vallere in the census section n. 2295, is near the Po river. The four rivers Po, Dora Riparia, Stura di Lanzo and Sangone can be recognized, respectively, in the east side near the hill, through Turin centre, in the northern part and in proximity to Turin's southern boundary.

In this work the causes of air temperature differences in the city of Turin are analysed considering also the different urban contexts in the census parcels in which the weather stations are localized.

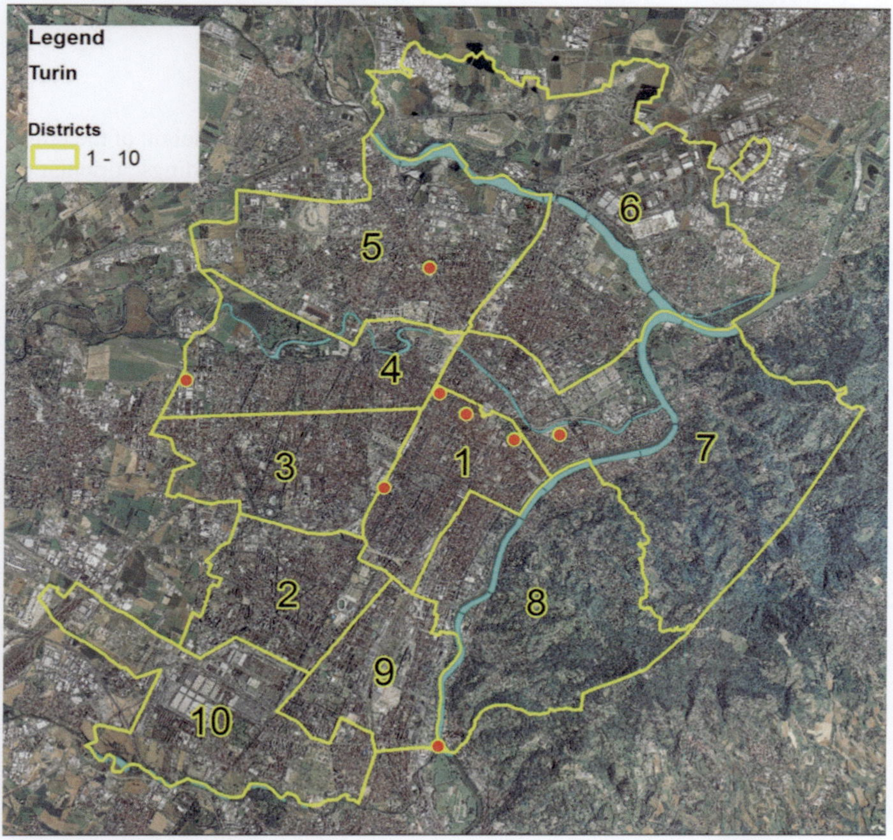

Fig. 11.3 The eight weather stations in the city of Turin with the representation of the urban context in which they are localized and the relative district (Source: author's elaboration)

11.4 Instruments and Methods

The aim of this work is to analyse how the built environment can influence the microclimate of different urban contexts. This analysis is focused on the census parcels, in which the weather stations of the city of Turin are localized (Fig. 11.2).

The variables analysed to characterize the built environment can be summarized in an urban morphology factor (*U*), a "solar exposure" (*P*) factor and the albedo coefficient (*A*) of outdoor materials.

The urban morphology characteristics (*U*) represent the contribution of the form of the built environment affecting the outdoor air temperature variation. The urban morphology is influenced mainly by the following variables: the buildings' density (*BD*) and the average buildings' distance (*W*) or the buildings' coverage ratio (*BCR*) and the aspect ratio (height to buildings' distance, *H*/*W*).

The urban morphology factor (U) can be expressed by

$$U = BD / W = BCR \cdot H / W[-]. \tag{11.1}$$

The solar exposure (P) is the function of the ratio between the height of the building and the height of the surroundings (H/H_{avg}), the main buildings' orientation (MBO) and the main orientation of the streets (MOS)

$$P = H / H_{avg} \cdot MBO \cdot MOS[-]. \tag{11.2}$$

The optimal orientation for buildings and streets for solar exposure is the east-west direction; then different values have been assigned to the *MBO* and *MOS* as shown in Fig. 11.4.

For the calculation of the albedo coefficient A, the correlations founded in literature have been used (Liang 2000 and Liang et al. 2002) using ASTER images. The sensor Advanced Spaceborne Thermal Emission and Reflection Radiometer (ASTER) is a satellite remote sensor providing high-resolution images of the earth in 14 different wavelengths on the electromagnetic spectrum.

ASTER images, in the Tagged Image File Format (TIFF), contain information relating to the radiance measured according to the wavelength of the sensor and expressed in the form of digital numbers (DN). To convert the DN in spectral radiance and top of atmosphere (TOA) radiance, specific correlations and coefficients can be used (Ghulam 2009; Banerjee et al. 2014).

From the spectral radiance and TOA radiance, it is possible to apply the equations of Liang, in order to calculate the albedo coefficient for all the wavelength bands of ASTER sensor. In particular, in this work, the data have been proceeded according to the Liang correlations for near-infrared ASTER albedo coefficient, using the satellite images relating to VNIR 1–2–3 bands (visible and near infrared) and 4–5–6–7–8–9 SWIR (shortwave infrared). The images of the ASTER sensor were provided by CSI (Consortium for Information System of Piedmont Region) and are related to the acquisition of July 22nd, 2004.

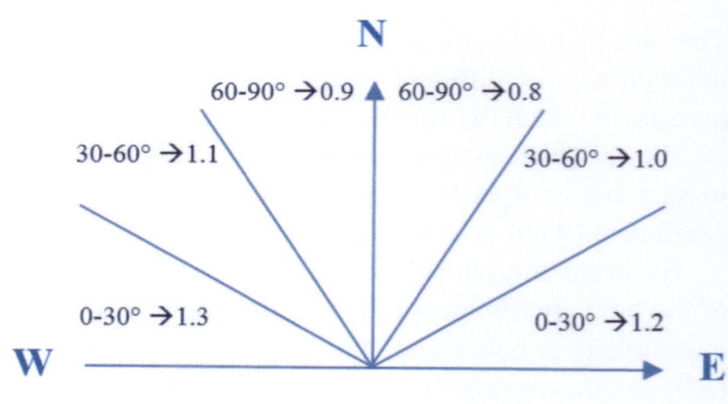

Fig. 11.4 Values for the main orientation of streets and buildings (MOS, MBO) (Source: author's elaboration)

Finally, each single urban variable is analysed with its average value considering its relative census section. For each census section, a global value G is then defined:

$$G(U,P,A)[-] \tag{11.3}$$

considering the variables of "urban morphology factor" (*U*), "solar exposure" (*P*) factor and albedo coefficient (*A*).

In this work, the urban variables of five different census sections are compared considering the average monthly air temperatures registered by the weather stations in Turin.

11.5 Results and Discussion

The city of Turin is localized in the Po Valley, and it is surrounded by beautiful mountains and hills. Turin is an important nineteenth-century city with historical buildings, beautiful squares and outdoor spaces but also industrial zones.

The area of the census sections in the city of Turin is not uniform due to districts with very large census sections as the hilly area (districts 7 and 8) and the industrial areas (districts 6 and 10) as shown in Figs. 11.2 and 11.3. The average size of the sections of the Turin census is about 35,000 m^2, with 17% of roads and 19% of built area (which becomes 23% if only the blocks of buildings are considered). Among this analysis, the census sections that are closest to these average values are the weather stations of Reiss Romoli (census section n.1886 in the northern area of Turin) and Buon Pastore (census section n.152 that is now no longer active). The characteristics of the census sections analysed with the weather stations are reported in Table 11.1.

Table 11.1 Characteristics of the census sections with the analysed weather stations

Turin's weather station	Weather station ARPA number	Census section number	Census section area (m^2)	Building area (m^2)	Building volume (m^3)	Streets area (m^2)	Blocks of buildings area (m^2)
via della Consolata	3447	14	8132	3529	59,535	2599	5533
Giardini Reali	446	66	79,070	565	6234	29,965	49,105
Reiss Romoli	3869	1886	52,790	16,866	210,907	9106	43,683
Vallere	249	2295	316,919	42,913	316,778	70,213	246,706
Alenia	4294	2414	405,096	146,901	1,686,958	39,184	365,912
Buon Pastore	153	152	52,756	13,963	165,401	8566	44,190
Italgas	145	960	130,323	23,165	172,352	40,245	90,079
Politecnico di Turin	–	773	183,878	41,534	418,532	97,665	86,214
Turin (average)		**1–3850**	**33,773**	**6426**	**83,008**	**5704**	**28,069**

In Figs. 11.5 and 11.6, the values of the "urban morphology factor" (*U*) and "solar exposure" (*P*) factor are represented for each census section of the city of Turin. From Fig. 11.5, the high buildings' density zone in the centre, the hilly part of the city in the south-east side and the industrial zones in the northern and southern parts can be recognized.

From Fig. 11.6, the solar exposure factor (*P*) is represented. The differences depend also by the orientation and the dimension of the census sections.

In Figs. 11.7 and 11.8, the albedo coefficient (*A*) is represented, respectively, with punctual values and average values for the census parcels. From Fig. 11.7, natural elements as the rivers and the hilly zone can be noticed; also the industrial roofs in the northern and southern part of the city can be noticed.

Finally, in Fig. 11.8, the average value of the albedo coefficient was calculated for each census parcel. Higher values can be noticed in the central zone of the city and along the rivers; instead, lower values of albedo coefficient characterize the industrial zones with lighter colour roofs.

Fig. 11.5 The urban morphology factor U for the census sections of the city of Turin

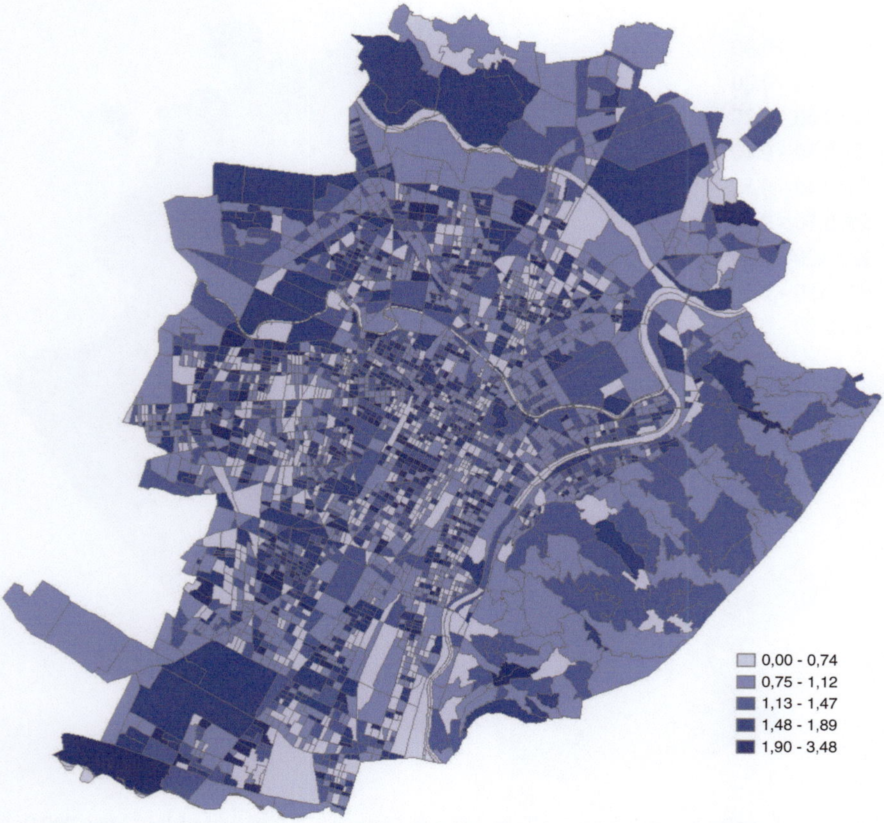

Fig. 11.6 The solar exposure factor P for the census sections of the city of Turin

In Table 11.2, the main characteristics about the urban form (*U*), solar exposure (*P*) and albedo coefficient (*A*) of the census parcels analysed are reported (with the ARPA weather stations). In particular, the high buildings' density (*BD*) values can be noticed for the areas with the weather stations of via della Consolata, Reiss Romoli, Alenia, Buon Pastore and Politecnico di Torino; lower values of *BD* can be observed in Giardini Reali and Vallere which are localized in parks and Vallere also near the Po river; finally, the Italgas weather station is in an ex-industrial zone.

The aspect ratio (*H/W*) is higher also for the high-density central zone, and, about the orientation of streets (*MOS*) and buildings (*MBO*), the main orientation is south, south-west and south-east.

Starting from these urban data, the aim of this work is to derive a model for the outdoor air temperature considering the average monthly data. This model could allow understanding, at the first design-planning phase, the effect of the materials, the geometric buildings' forms, the characteristics of outdoor spaces and the solar exposition, on the performance of air temperatures; the model could be used as a

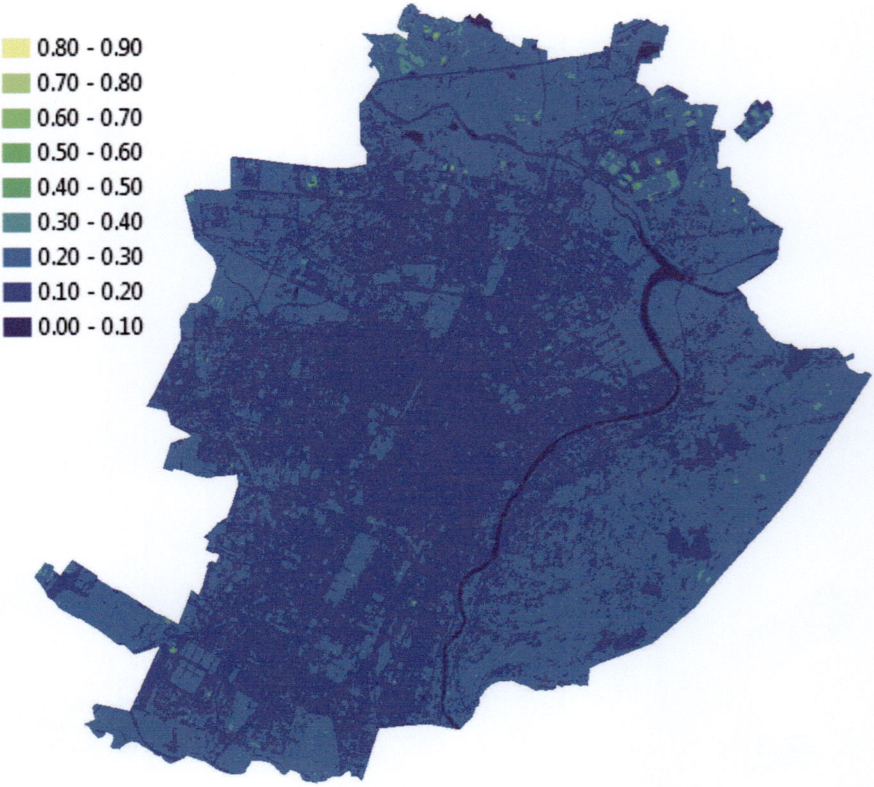

Fig. 11.7 The albedo coefficient A in Turin (obtained from ASTER images of July 22nd, 2004)

tool to improve environmental sustainability and urban liveability in urban planning.

For a good design tool, the model should evaluate the average monthly air temperature for a prediction in all seasons, especially during wintertime and summertime.

A multiple linear regression analysis correlates the urban variables $G(U, P, A)$ to the average air temperatures considering also a typical monthly air temperature behaviour for Turin (ΔT_m)

$$\begin{aligned} T_{air} &= f(GUPA) = f('' \, T_m, BCR, H/W, MOS, H/H_{\text{aver}}, MBO, A) \\ &= (\alpha_1 '' \, T_m) + (\alpha_2 \cdot BCR) + (\alpha_3 \cdot H/W) + (\alpha_4 \cdot MOS) + (\alpha_5 \cdot H/H_{\text{aver}}) + (\alpha_6 \cdot MBO) \\ &\quad + (\alpha_7 \cdot A)[^\circ\text{C}]. \end{aligned} \tag{11.4}$$

where ΔT_m is the monthly variation of the air temperatures considering weather data from 2006 to 2014 averaged on the actual functioning five ARPA weather stations in Turin: via della Consolata, Giardini Reali, Reiss Romoli, Vallere and Alenia.

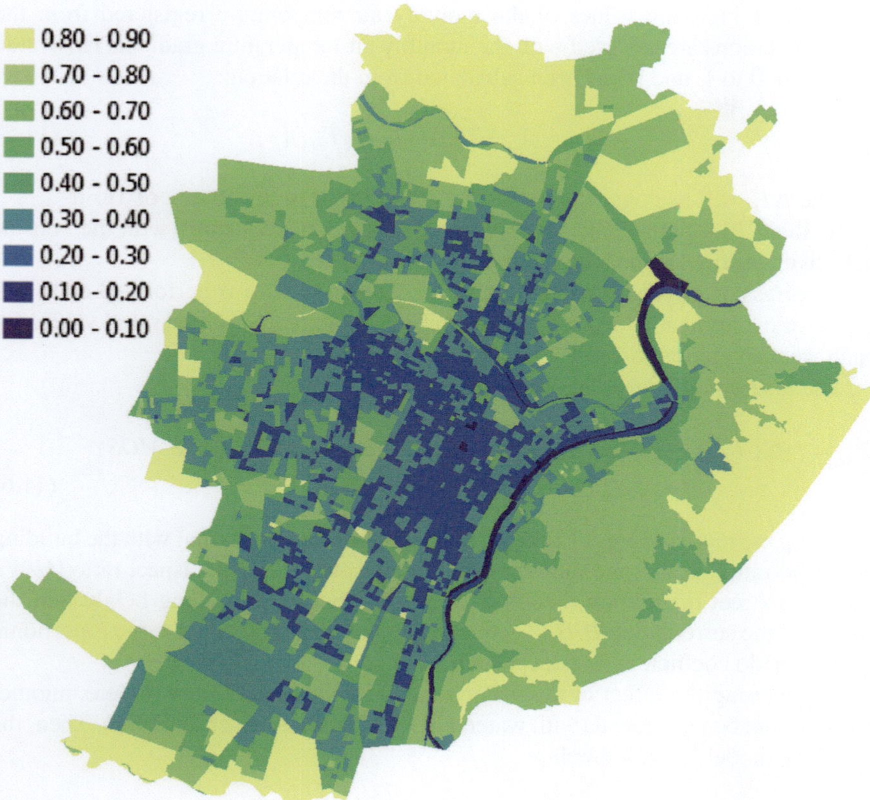

Fig. 11.8 The average value of albedo coefficient A for the census sections of Turin

Table 11.2 Characteristics of the census sections with the analysed weather stations

Turin's weather station	BCR (m^2/m^2)	BD (m^3/m^2)	H (m)	H/W (m/m)	H/H_{avg} (m/m)	MBO (–)	MOS (–)	U (–)	P (–)	A (–)
via della Consolata	0.43	7.3	17.6	0.80	1.0	1.3	1.3	0.35	1.8	0.15
Giardini Reali	0.01	0.1	13.1	0.18	1.1	1.1	1.1	0.00	1.3	0.21
Reiss Romoli	0.32	4.0	18.2	0.56	1.0	0.9	1.1	0.18	0.9	0.20
Vallere	0.14	1.0	8.4	0.58	0.8	0.8	0.8	0.08	0.5	0.20
Alenia	0.36	4.2	13.5	0.47	1.4	0.8	0.8	0.17	0.9	0.19
Buon Pastore	0.26	3.1	15.9	0.66	1.1	1.3	1.3	0.17	1.8	0.17
Italgas	0.18	1.3	9.8	0.19	0.9	1.3	1.3	0.03	1.5	0.18
Politecnico di Torino	0.39	5.8	17.9	0.54	0.9	1.3	0.8	0.21	0.9	0.17
Turin (average)	**0.34**	**4.96**	**17.85**	**0.70**	**1.0**	**1.1**	**1.05**	**0.26**	**1.17**	**0.18**

In Table 11.3, the values of the monthly air temperature registered from the weather stations are reported with the monthly air temperature gradient. This value goes from 0 to 1, and it has been calculated from the relation:

$$\Delta T_m = (T_m - T_{\min}) / (T_{\max} - T_{\min})[-] \tag{11.5}$$

and the ΔT_m will be considered uniformly distributed in the all city of Turin.

In Table 11.4 the average characteristics of the census section with the considered five weather stations are reported.

A regression analysis of the monthly air temperatures was performed considering the most influential urban variables in Eq. 11.4 to reach the measured values for each weather station in Table 11.3.

The resulting model is the following:

$$T_{air} = f(GUPA) = (22.75\cdot\text{''}\ T_m) + (3.31\cdot BCR) + 0.10\cdot H/W) + (2.38\cdot MOS) + (0.55\cdot H/H_{\text{aver}}) + (-0.44\cdot A)[°C] \tag{11.6}$$

Using the model of Eq. 11.6, the air temperature is proportional with the building coverage ratio (*BCR*) (and then the building density (*BD*)), the aspect ratio (*H/W*), the main street orientation (*MOS*) and the ratio between building height and the height of the surrounding H/H_{aver}, while the air temperature is inversely proportional to the albedo coefficient *A*, as expected.

Considering the effect of the water near the weather station of Vallere, another variable can be introduced (with water, $H_2O = 1$; without water, $H_2O = 0$). Then, the resulting model is the following:

$$T_{air} = f(GUPA) = (22.76\cdot\text{''}\ T_m) + (2.19\cdot BCR) + (0.99\cdot H/W) + (2.09\cdot MOS) + (0.74\cdot H/H_{\text{aver}}) + (-0.56\cdot A) + (-0.53\cdot H_2 0)[°C]. \tag{11.7}$$

with the air temperature inversely proportional also to the presence of water, as expected.

Also the presence of vegetation was investigated, with a similar indicator as the one for the presence of water, but without better results.

As an example in Figs. 11.9 and 11.10, the results of the models (Eqs. 11.6 and 11.7) and the measured values of air temperatures are reported for the weather stations of via della Consolata and Vallere.

In Table 11.5, the relative errors of the models used are reported for all the weather stations considered. As it is possible to notice, the relative errors are under 10% with both the models considering (with Eq. 11.7) or not (with Eq. 11.6) the presence of water.

In Figs. 11.11, 11.12, 11.13, 11.14, 11.6, 11.17 and 11.18, the evaluations of the outdoor air temperatures in the city of Turin are represented with the model of Eqs. 11.6 and 11.7 as a function of the urban morphology, the solar factor, the albedo

Table 11.3 Monthly air temperatures (2005–2015) and the relative value of monthly gradients ΔT_m

Month	4	5	6	7	8	9	10	11	12	1	2	3
Via della Consolata	21.30	18.30	23.50	28.30	25.30	22.40	10.00	11.00	6.20	5.00	8.00	13.80
Giardini Reali	19.50	16.70	21.90	27.30	23.90	21.30	9.50	7.70	3.00	3.20	7.00	11.20
Reiss Romoli	20.80	17.50	23.00	27.20	24.50	21.60	9.10	9.90	5.30	3.80	7.60	12.60
Vallere	19.40	16.30	21.20	26.90	23.70	20.60	9.60	7.30	2.10	2.50	6.60	10.30
Alenia	20.50	17.30	22.50	27.10	24.10	21.10	8.90	10.60	5.30	4.00	7.10	12.80
T_m [°C]	20.30	17.22	22.42	27.36	24.30	21.40	9.42	9.30	4.38	3.70	7.26	12.14
Δ *Tm* [–]	**0.65**	**0.56**	**0.80**	**1.00**	**0.88**	**0.75**	**0.28**	**0.22**	**0.02**	**0.00**	**0.14**	**0.34**

Table 11.4 Statistical data of the census parcel with the five weather stations

	BCR (m^2/m^2)	BD (m^3/m^2)	BH (m)	H/W (m/m)	H/H_{avg} (m/m)	MBO (−)	MOS (−)	U (−)	P (−)	A (−)
Average	0.25	3.32	14.16	0.52	1.06	0.98	1.02	0.16	1.08	0.19
Standard deviation	0.17	2.86	3.97	0.22	0.22	0.22	0.22	0.13	0.49	0.02

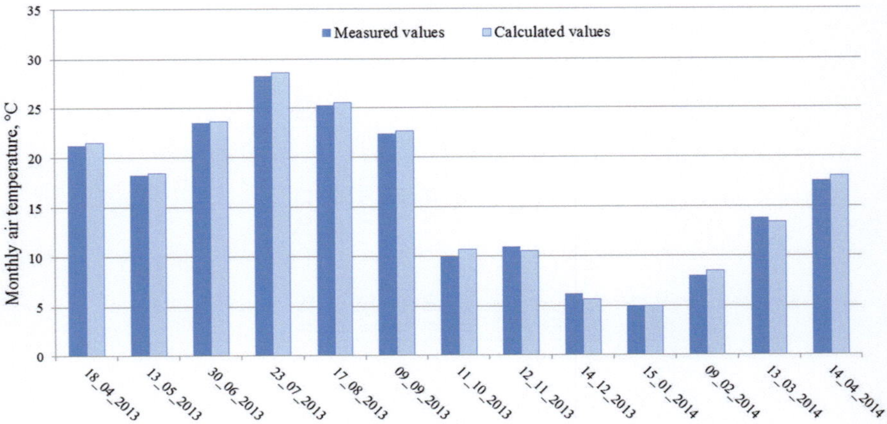

Fig. 11.9 Comparison between measured air temperatures with the weather station of via della Consolata and the calculated air temperatures as a function of *G*(*UPA*) with the model (Eq. 11.6)

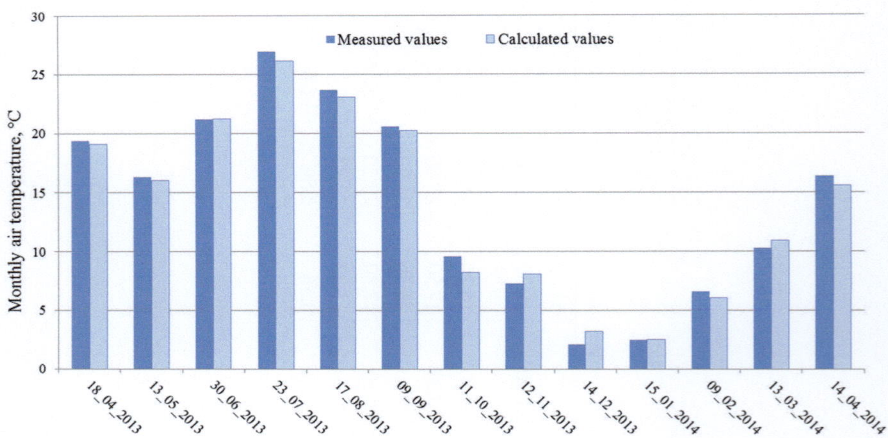

Fig. 11.10 Comparison between the measured air temperatures with the weather station of Vallere (with Po river) and the calculated air temperatures as a function of G(UPA) with the model (Eq. 11.7)

Table 11.5 The relative errors between the measured air temperatures and the calculated values with the models (Eqs. 11.6 and 11.7) for the census sections with the ARPA weather stations

Weather station	Relative error – %	
	Model Eq. 11.6	Model Eq. 11.7
Via della Consolata	3.3	3.4
Giardini Reali	4.9	4.9
Reiss Romoli	2.6	2.8
Vallere	9.5	8.2
Alenia	4.1	4.1

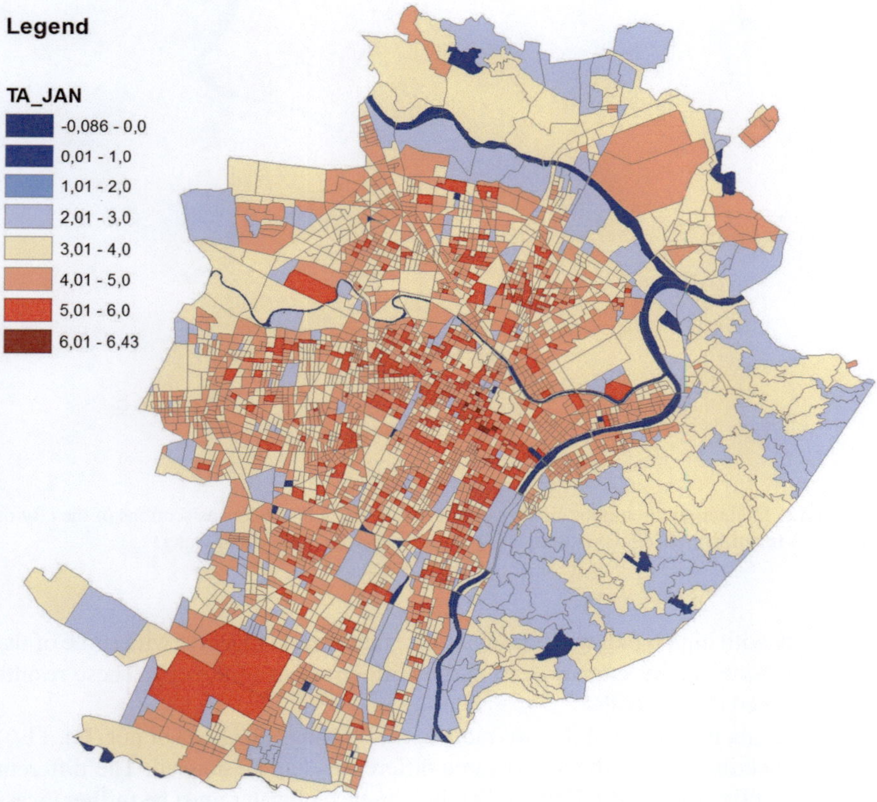

Fig. 11.11 Evaluation of outdoor air temperatures in January in the census sections of the city of Turin as a function of $G(UPA)$ with Eq. 11.6

coefficient and the presence of water for different months during the heating and cooling seasons.

As it is possible to note, the areas near the rivers are the most cold zones with some of the areas in the hill (south-east) or near the parks. The areas near the centre

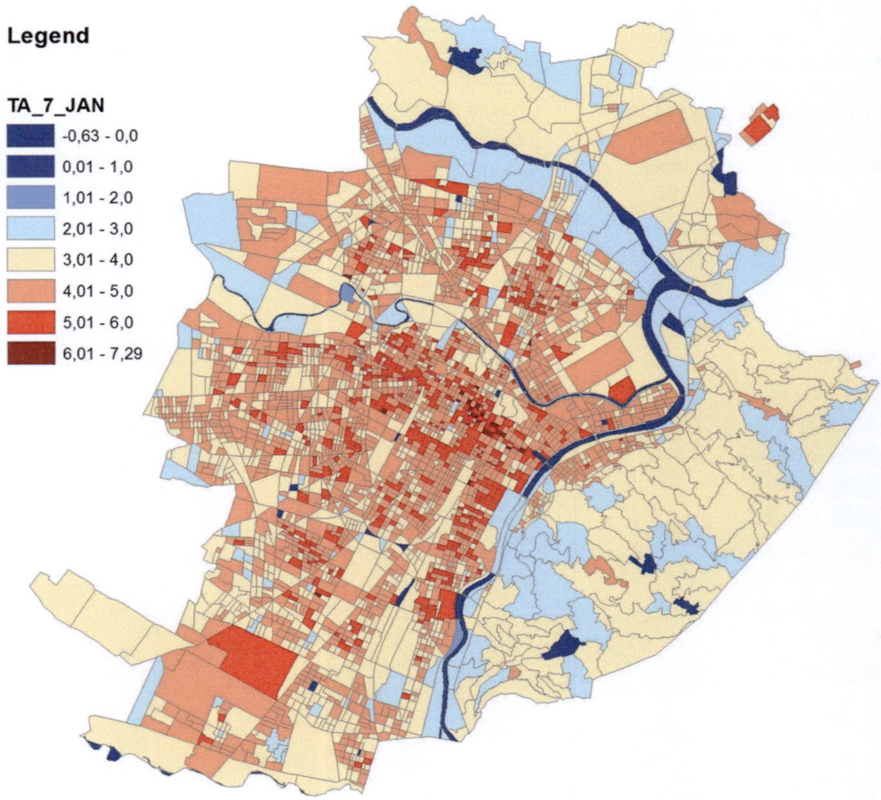

Fig. 11.12 Evaluation of outdoor air temperatures in January in the census sections of the city of Turin as a function of $G(UPA)$ with Eq. 11.7 (considering the presence of water)

of the city with high buildings' density are the more warm together with some of the industrial zones in the southern and in the northern parts of the city. These results are confirmed in all months.

Comparing the results of the two models considering (Eq. 11.7) or not (Eq. 11.6) the presence of water in the areas, some differences can be noticed. The different influence of the aspect ratio H/W and of the presence of water must be further investigated in future works with more weather stations inside the city of Turin and comparing the results of the model with other data as the Landsat satellite images; also the presence of vegetation should be analysed.

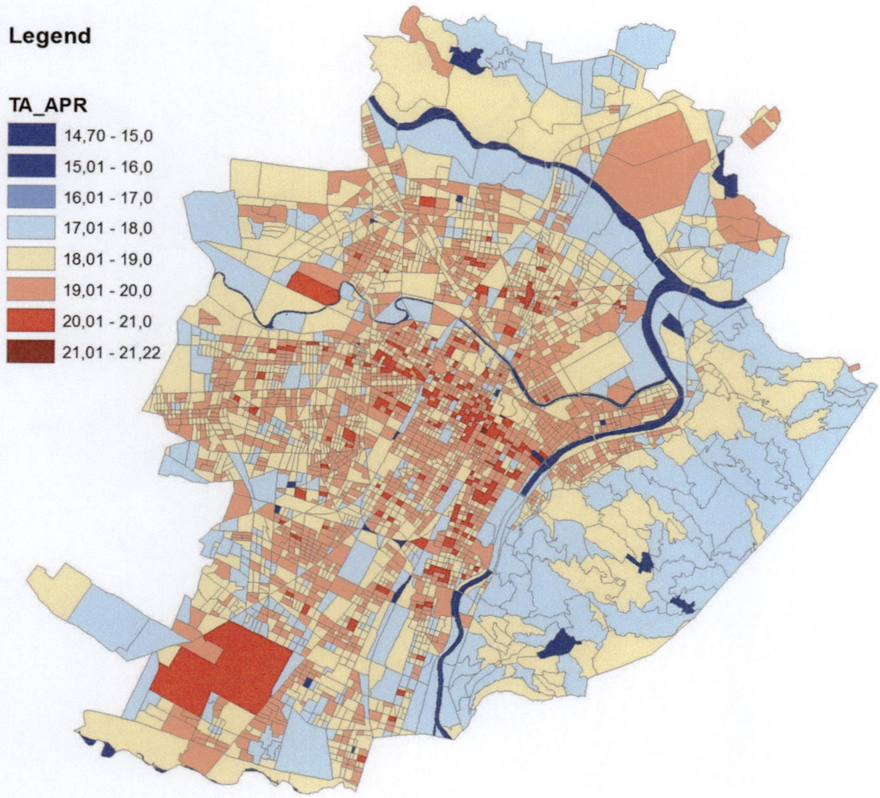

Fig. 11.13 Evaluation of outdoor air temperatures in April in the census sections of the city of Turin as a function of $G(UPA)$ with Eq. 11.6

11.6 Conclusion

In this work, the analysis of urban microclimate, the effect of urban variables and the presence of water and green surfaces have been conducted on the city of Turin, using information from the following sources: ARPA Piemonte weather stations, ASTER satellite data and urban variables from the Technical Map of the Metropolitan City of Turin.

The data acquired by weather stations are used to estimate the average monthly air temperatures measured in the areas with the weather stations in Turin; from the ASTER data, the albedo coefficient values were obtained for each point of the city and for each census sections of Turin. The Technical Map of the Metropolitan City of Turin has been used to evaluate all geometrical characteristics of the built heritage of the city of Turin to calculate the urban morphology factor U and the solar exposure P.

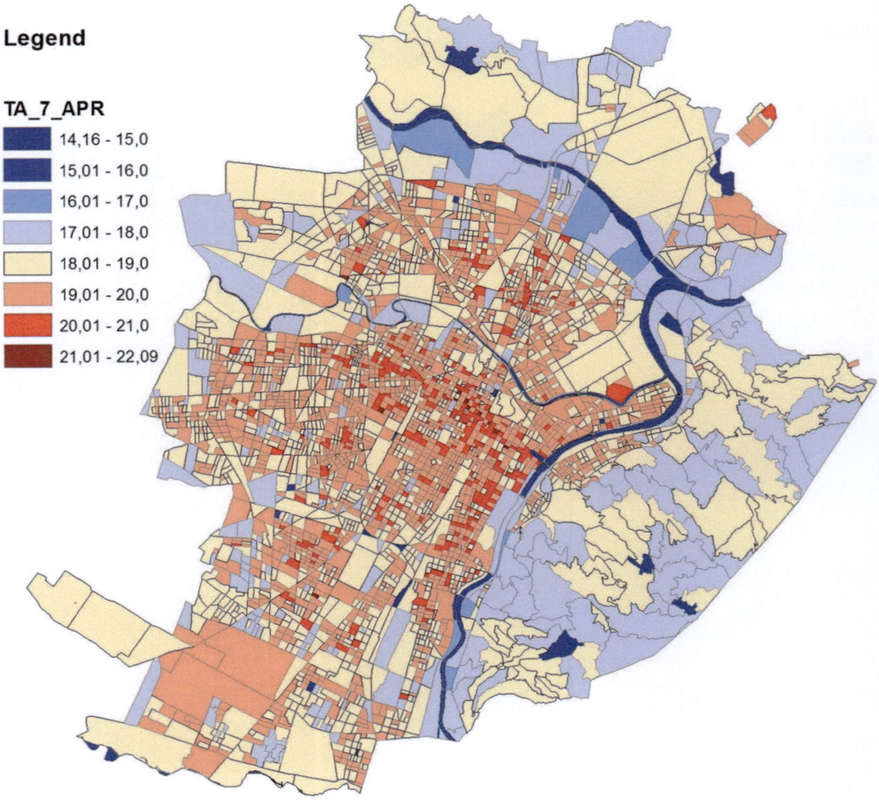

Fig. 11.14 Evaluation of outdoor air temperatures in April in the census sections of the city of Turin as a function of *G*(*UPA*) with Eq. 11.7 (considering the presence of water)

All the above information have been used to formulate two simplified models for the calculation of the outdoor air temperature: the first model calculates the air temperature taking into account the urban planning variables and the second model considering also the presence or the absence of the water.

The application of the model consents to map the air temperatures in the city of Turin for each month, according to the urban variables and the albedo coefficient of the outdoor spaces to understand the air temperature variations.

The model can be also applied in different design phases of the urban planning; it can be applied in the predesign phase to see how the project can influence the outside air temperature and then the liveability of the outdoor spaces in all the seasons. Moreover, the model can also be used to check how the different urban vari-

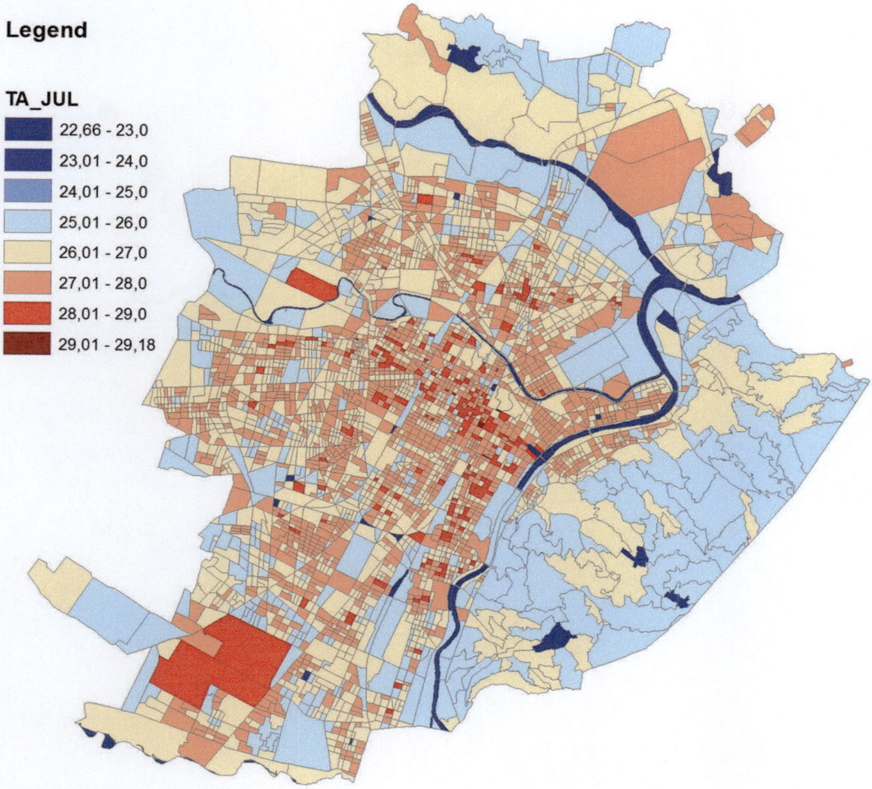

Fig. 11.15 Evaluation of outdoor air temperatures in July in the census sections of the city of Turin as a function of $G(UPA)$ with Eq. 11.6

ables can influence the outdoor air temperatures and to evaluate how to improve the liveability of the urban environment with interventions that can mitigate the microclimate at blocks of building scale.

Future works will improve the models considering more areas around Turin with weather stations and comparing also the results of the models with other satellite images as the Landsat ones.

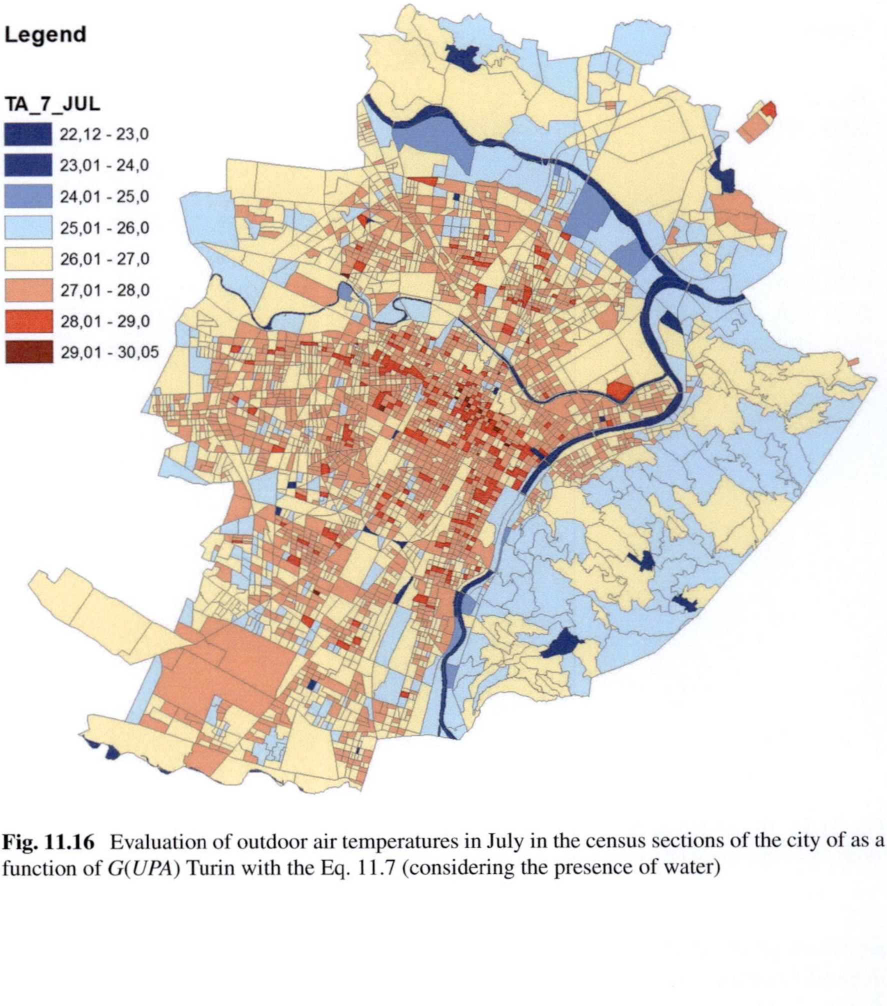

Fig. 11.16 Evaluation of outdoor air temperatures in July in the census sections of the city of as a function of *G*(*UPA*) Turin with the Eq. 11.7 (considering the presence of water)

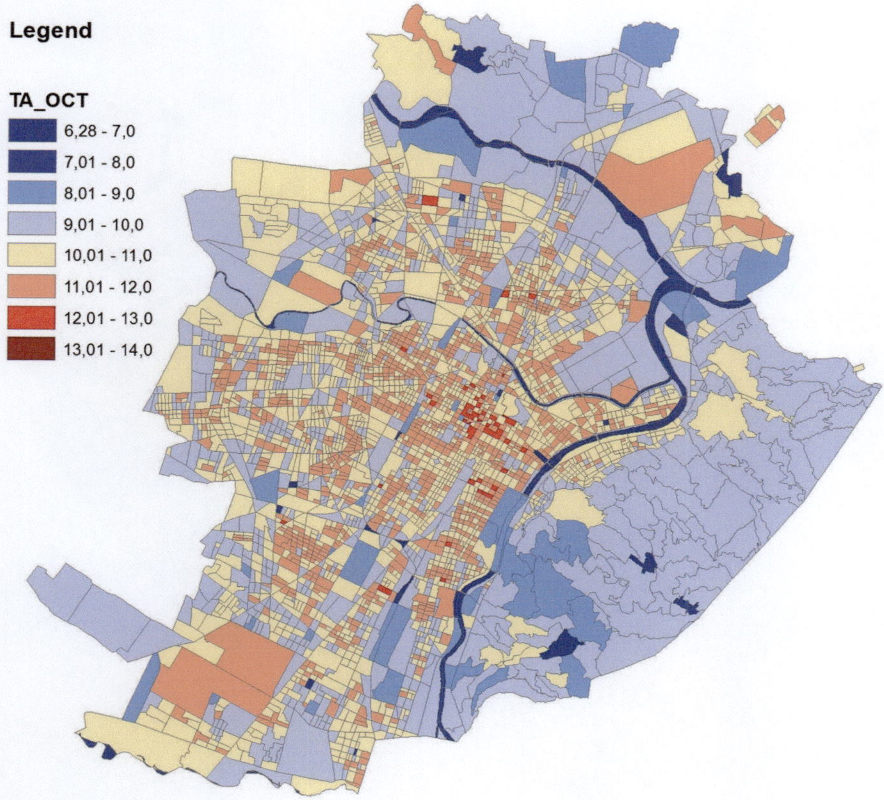

Fig. 11.17 Evaluation of outdoor air temperatures in October in the census sections of the city of Turin as a function of *G*(*UPA*) with Eq. 11.6

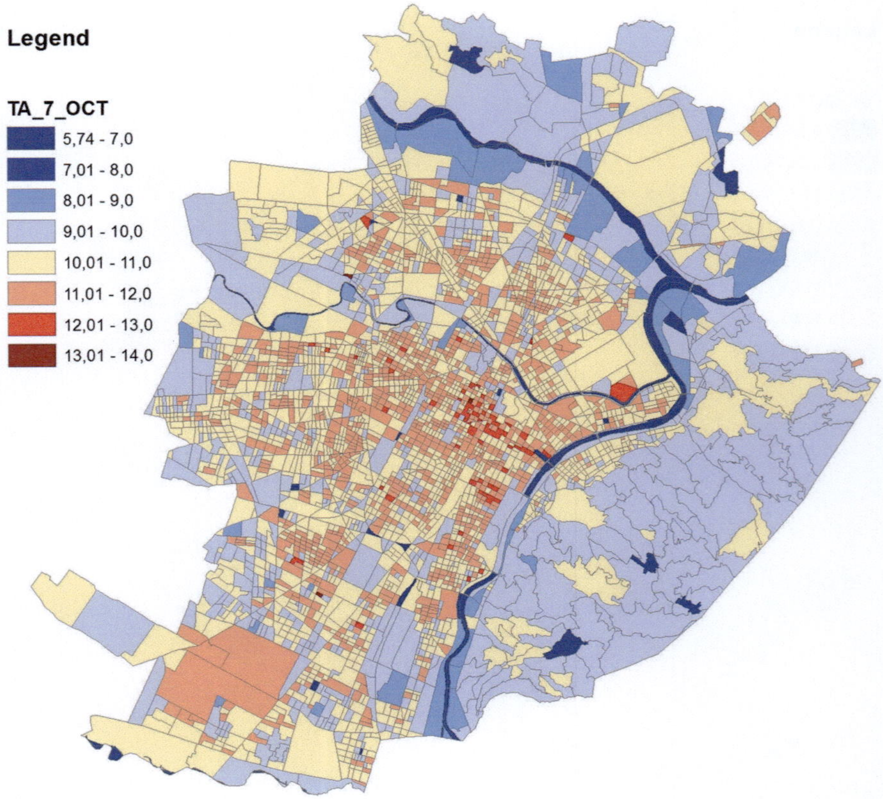

Fig. 11.18 Evaluation of outdoor air temperatures in October in the census sections of the city of Turin as a function of G(UPA) with Eq. 11.7 (considering the presence of water)

Acknowledgments A special thanks to the Master Degree students Gualtiero Ceretti, Marta Carozza, Daria Carta, José Correa Martin and Marta Repice for their support in the analysis of urban variables and in the evaluation of the albedo coefficients and finally to Chantal Diegoli (CSI Piemonte) for sharing ASTER spatial data.

References

Allegrini J et al (2014) Influence of the urban microclimate in street canyons on the energy demand for space cooling and heating of buildings. Energy and Buildings 55:823–832

Aniello C et al (1995) Mapping micro-urban heat islands using Landsat TM and a GIS. Comput Geosci 21(8):965–969

Balogun AA et al (2014) Effect of tree-shading on energy demand of two similar buildings. Energy and Buildings 81:305–315

Banerjee K, Panda S, Jain M K, Jeyaseelan A T, Sharma R Kr (2014) Comparison of aster thermal bands and feature identification using advance spectroscopic techniques. Int J Innov Sci Res 7(1):11–18
Berger T et al (2015) Impacts of climate change upon cooling and heating energy demand of office buildings in Vienna, AustriaTania. Energy and Buildings 80:517–530
Carnielo E, Zinzi M (2014) Optical and thermal characterisation of cool asphalts to mitigate urban temperatures and building cooling demand. Build Environ 60:56–65
Chun B, Guldmann J-M (2014) Spatial statistical analysis and simulation of the urban heat island in high-density central cities. Landsc Urban Plan 125:76–88
Coseo P, Larsen L (2015) Cooling the heat island in compact urban environments: the effectiveness of Chicago's Green Alley Program. Procedia Engineering 118:691–710
Dabaieh M et al (2015) Reducing cooling demands in a hot dry climate: a simulation study for non-insulated passive cool roof thermal performance in residential buildings. Energy and Buildings 89:142–152
Debbage N, Shepherd JM (2015) The urban heat island effect and city contiguity. Comput Environ Urban Syst 54:181–194
Delmastro C, Mutani G, Pastorelli M, Vicentini G (2015a) Urban morphology and energy consumption in Italian residential buildings. 4th International Conference on Renewable Energy Research and Applications ICRERA-2015, Palermo, 22–25 November 2015
Delmastro C, Mutani G, Schranz L, Vicentini G (2015b) The role of urban form and socio-economic variables for estimating the building energy savings potential at the urban scale. Int J Heat Technol 33(4):91–100
Effat HA, Hassan OAK (2014) Change detection of urban heat islands and some related parameters using multi-temporal Landsat images; a case study for Cairo city, Egypt. Urban Climate 10:171–188
Feng H et al (2014) Using land use change trajectories to quantify the effects of urbanization on urban heat island. Adv Space Res 53:463–473
Fernàndez FJ et al (2015) Optimal location of green zones in metropolitan areas to control the urban heat island. J Comput Appl Math 289:412–425
Ghulam A (2009) How to calculate reflectance and temperature using ASTER data, Center for Environmental Sciences at Saint Louis University http://www.pancroma.com/downloads/ASTER%20Temperature%20and%20Reflectance.pdf. Accessed 29 June 2016
Gracik S et al (2015) Effect of urban neighborhoods on the performance of building cooling systems. Build Environ 90:15–29
Houet T, Pigeon G (2011) Mapping urban climate zones and quantifying climate behaviors e An application on Toulouse urban area (France). Environ Pollut 159:2180–2192
Iino A, Hoyano A (1996) Development of a method to predict the heat island potential using remote sensing and GIS data. Energy and Buildings 23:199–205
Ivajnsic D, Kaligaric M, Ziberna I (2014) Geographically weighted regression of the urban heat island of a small city. Appl Geogr 53:341–353
Jusuf S K et al. (2007) The influence of land use on the urban heat island in Singapore, Habitat Int 31:232–242
Kaloustian N, Diab Y (2015) Effects of urbanization on the urban heat island in Beirut. Urban Clim 14(Part 2):154–165
Kolokotroni M et al (2014) A validated methodology for the prediction of heating and cooling energy demand for buildings within the urban heat island: case-study of London. Sol Energy 84:2246–2255
Kong F et al (2014) Effects of spatial pattern of greenspace on urban cooling in a large metropolitan area of eastern China. Landsc Urban Plan 128:35–47
Li J et al (2009) Remote sensing evaluation of urban heat island and its spatial pattern of the Shanghai metropolitan area, China. Ecol Complex 6:413–420
Li J et al (2011) Impacts of landscape structure on surface urban heat islands: a case study of Shanghai, China. Remote Sens Environ 115:3249–3263

Li Y-Y, Zhang H, Kainz W (2012) Monitoring patterns of urban heat islands of the fast-growing Shanghai metropolis, China: using time-series of Landsat TM/ETM+ data. Int J Appl Earth Obs Geoinf 19:127138

Liang S (2000) Narrowband to broadband conversions of land surface albedo I Algorithms. Remote Sens Environ 76:213–238

Liang S, Shuey CJ, Russ AL, Fang H, Chen M, Walthall C, Daughtry CST, Hunt RJ (2002) Narrowband to broadband conversions of land surface albedo: II. Validation Remote Sens Environ 84:25–41

Liu J et al (2015) Numerical evaluation of the local weather data impacts on cooling energy use of buildings in an urban area. Procedia Engineering 121:381–388

Mallick J, Rahman A, Singh CK (2013) Modeling urban heat islands in heterogeneous land surface and its correlation with impervious surface area by using night-time ASTER satellite data in highly urbanizing city, Delhi-India. Adv Space Res 52(4):639–655

Mazzotta A, Mutani G (2015) Environmental high performance urban land cover: experimentations in Urban Barriera. 6th International Building Physics Conference, Turin 14–17th June 2015, Energy Procedia, Doi:10.1016/j.egypro.2015.11.059

Merbitz H et al (2012) GIS-based identification of spatial variables enhancing heat and poor air quality in urban areas. Appl Geogr 33:94–106

Mirzaei PA (2015) Recent challenges in modeling of urban heat island. Sustainable Cities and Society 19:200–206

Misni A (2015) The effect of building construction and human factors in cooling energy use. Procedia – Soc Behav Sci 202:373–381

Morris KI et al (2015) Computational study of urban heat island of Putrajaya, Malaysia. Sustainable Cities and Society 19:359–372

Mutani G (2016) Urban planning for the liveability and thermal comfort of outdoor spaces. 11th conference on sustainable of energy, water and environmental systems, SDEWES2016.0442. Lisbon, pp 1–15, ISSN: 1847–7178

Mutani G, Gamba A, Maio S (2016) Space heating energy consumption and urban form. The case study of residential buildings in Turin (Italy). 11th conference on sustainable of energy, water and environmental systems, SDEWES2016.0441. Lisbon, pp 1–17. ISSN: 1847–7178

Onischi A et al (2010) Evaluating the potential for urban heat-island mitigation by greening parking lots. Urban For Urban Green 9:323–332

Peron F et al (2015) Urban Heat Island (UHI) mitigating strategies: a case-based comparative analysis. Sustainable Cities and Society 19:222–235

Pisello AL (2015) Thermal-energy analysis of roof cool clay tiles for application in historic buildings and cities. Sustain Cities Soc 19:271–280

Pisello AL, Cotana F (2014) The thermal effect of an innovative cool roof on residential buildings in Italy: results from two years of continuous monitoring. Energy and Buildings 69:154–164

Pisello AL, Cotana F (2015) Thermal-energy and environmental impact of cool clay tiles for residential buildings in Italy. Procedia Eng 118:530–537

Qin Y (2015) A review on the development of cool pavements to mitigate urban heat island effect. Renew Sust Energ Rev 52:445–459

Quan SJ et al (2015) A GIS-based energy balance modeling system for urban solar buildings. Energy Procedia 75:2946–2952

Raghavan K et al (2015) Influence of urban a reason environment: special reference to building materials and temperature anomalies using geospatial technology. Sustainable Cities and Society 19:349–358

Sailor DJ (2014) Risks of summertime extreme thermal conditions in buildings as a result of climate change and exacerbation of urban heat islands. Build Environ 78:81–88

Salata F et al (2015) How high albedo and traditional buildings' materials and vegetation affect the quality of urban microclimate. A case study. Energy Build 99:32–49

Santamouris M (2014a) Cooling the cities – a review of reflective and green roof mitigation technologies to fight heat island and improve comfort in urban environments. Sol Energy 103:682–703
Santamouris M (2014b) On the energy impact of urban heat island and global warming on buildings. Energy and Buildings 82:100–113
Santamouris M et al (2015) On the impact of urban heat island and global warming on the power demand and electricity consumption of buildings – a review. Energy Build 98:119–124
Schwartz N et al (2012) Relationship of land surface and air temperatures and its implications for quantifying urban heat island indicators – an application for the city of Leipzig (Germany). Ecol Indic 18:693–704
Shahidan MF et al (2015) An evaluation of outdoor and building environment cooling achieved through combination modification of trees with ground materials. Build Environ 58:245–257
Stathopoulou M, Cartalis C (2009) Downscaling AVHRR land surface temperatures for improved surface urban heat island intensity estimation. Remote Sens Environ 113:2592–2605
Theophilou MK, Serghiedes D (2015) Estimating the characteristics of the Urban Heat Island effect in Nicosia, Cyprus, using multiyear urban and rural climatic data and analysis. Energy Build 108:137–144, 157
Touchaei AG, Wang Y (2015) Characterizing urban heat island in Montreal (Canada) – effect of urban morphology. Sustain Cities Soc 19:395–402
Unger J et al (2010) Comparison and generalisation of spatial patterns of the urban heat island based on normalized values. Phys Chem Earth 35:107–114
United Nations (2014) World Urbanization Prospects: the 2014 Revision, Highlights (ST/ESA/SER.A/352). Department of Economic and Social Affairs, Population Division, ISBN 978-92-1-123195-3
Weng Q, Lung D, Schubring J (2004) Estimation of land surface temperature–vegetation abundance relationship for urban heat island studies. Remote Sens Environ 89:467–483
Xu LY, Xie XD, Li S (2013) Correlation analysis of the urban heat island effect and the spatial and temporal distribution of atmospheric particulates using TM images in Beijing. Environ Pollut 178:102–114
Yang J, Wang ZH, Kaloush KE (2015) Environmental impacts of reflective materials: is high albedo a 'silver bullet' for mitigating urban heat island? Renew Sust Energ Rev 47:830–843
Yeo IA, Yoon SH, Yee JJ (2013) Development of an Environment and energy Geographical Information System (E-GIS) construction model to support environmentally friendly urban planning. Appl Energy 104:723–739
Zhang Y et al (2012) Study on urban heat island effect based on normalized difference vegetated index: a case study of Wuhan City. Procedia Environ Sci 13:574–581
Zhang H et al (2013) Analysis of land use/land cover change, population shift, and their effects on spatiotemporal patterns of urban heat islands in metropolitan Shanghai, China. Appl Geogr 44:121–133
Zhang B et al (2014) The cooling effect of urban green spaces as a contribution to energy-saving and emission-reduction: a case study in Beijing, China. Build Environ 76:37–43

Part III
Debate

The third part of the book focuses on utopias and visionary thinking (Palazzo and Diko) as an important contribution to social evolution and territorial visioning as recognized by the major literature on architecture, urban planning, and design. Ecological planning uses utopian visioning in problem-solving and adapting solutions for a balance among spatial, environmental, cultural, social, and economic values (i.e., in US tradition). Theoretical frameworks and methods are different design tools used for negotiation in complex situations, dealing with different scales and disciplinary bias, and design outcomes (Plowright).

Part III
Debate

Chapter 12
From the Past into the Future: The Utopian Roots of Ecological Planning in North America

Danilo Palazzo and Stephen K. Diko

Abstract In the 1960s, ecological planning gained prominence in North America as a way for managing the dynamic and complex relationships between human settlements and the natural environment. In subsequent years, ecological planning developed as a planning specialization in North America, assuming a prominent role in discussions on the environmental crisis that persisted during the period. We argue that ecological planning emerged from the evolution of three initial fields of studies and activities: (i) English landscape gardening that was transmuted into genuine American landscape architecture; (ii) the management of the vast public domain, particularly by some federal agencies; and (iii) the concept of regional planning as advocated by the Regional Planning Association of America and then during the New Deal. In this paper, we conduct a historical review of these three fields, that we call roots, and then argue that utopian thinking and its translation into utopian communities in the United States, starting from the end of seventeenth century, represent the fourth field that contributed to the formative processes of ecological planning in North America. Additionally, we show how the past, in terms of utopian thinking, provides planning with a framework to confront current and future urban challenges.

Keywords Ecological planning • Environment • Planning • North America • Utopia

12.1 Introduction

It was not until the 1960s that the term ecological planning in North America emerged as a dominant planning specialization to understand and confront the problems of human settlements. Before the United States' adoption of the National Environmental Policy Act (NEPA) of 1969, discussions on the interrelationships

D. Palazzo (✉) • S.K. Diko
School of Planning, University of Cincinnati, 45221 Cincinnati, OH, USA
e-mail: palazzdo@ucmail.uc.edu; dikosk@mail.uc.edu

R. Ingaramo, A. Voghera (eds.), *Topics and Methods for Urban and Landscape Design*, Urban and Landscape Perspectives 19,
DOI 10.1007/978-3-319-51535-9_12

between the natural and built environments and their management were infrequent if not missing entirely in most North American planning practice. The agency, NEPA, revitalized concerns for managing the relationships between the natural and built environments, while ecological planning surfaced in recognition of their interconnected complexity. Furthermore, ecological planning surfaced in response to the newly recognized need to plan and design cities and regions to optimize, respect, and preserve ecological forces, flows, and manifestations.

Subsequently, the 1960s and 1970s, with greater public awareness of increasing environmental crisis, served as a period of incubation and growth of ecological planning. As William M. Marsh notes, the environmental crisis "paved the way for stronger and broader environmental legislation at all levels of government" (Marsh 2005, 3). Although the role of ecology in planning had languished after World War II—with some exceptions such as the work of Hideo Sasaki (Walker and Simo 1994) or Garrett Eckbo (Eckbo 1950)—the crisis gave rise to another level of discourse and studies on the ecology or the environment and its role in planning and design. By this time, ecological planning had become "a way of directing or managing changes in the landscape so that human actions are in tune with natural processes" (Ndubisi 2002, 2). Furthermore, ecological planning was seen as a tool as well as a process for managing the relationships between human activities and the natural environment with emphasis on how people used land and other natural resources (Ndubisi 2002).

George Angus Hills, Philip H. Lewis, and Ian McHarg were the prominent personalities that employed ecologically conscious methods during this time (Belknap and Furtado 1967), significantly shaping the ecological planning field. They provided the technical elements and applications that paved the way for appreciating, understanding, analyzing, and producing plans and designs through ecological lenses to manage the interaction between the built and natural environments. However, these forerunners were far from inventing an entirely new field.

Apparently, ecological planning emerged from the culmination of the processes, practices, and notions that were already components of the North America's planning history. The main protagonists acted as catalyzers of an existing but often hidden past of ecological planning (Palazzo 1997). The past of ecological planning remained obscure even to those who contributed to its invention. During the seminal moments of ecological planning, attention to what came before was very weak. This is reasonable because the authors focused on pragmatic rather than theoretical and historical issues. The hidden past comprises of three main roots that evolved in a timeframe spanning from the early days of American independence to World War II, namely, the initiation of landscape architecture, the management of the vast public domain, and the philosophy of regional planning (Palazzo 2003).

In this paper, we briefly describe the three roots since they have been extensively discussed in previous works (Palazzo 1997, 2003) and then broadly explore a fourth root of ecological planning: utopian thinking. We argue that, like the other three roots, utopian thinking led to the consolidation of ecological planning as a planning specialization in North America. Although utopia is a contentious subject, we argue that it offers critical perspectives into the formative processes that shaped the theoretical and philosophical elements of ecological planning as a planning specialization in the 1960s and 1970s, which later became synonymous to environmental planning. The utopian notions of early European settlers in communitarian settlements and influential ideas of visionaries such as Ebenezer Howard (1850–1928), Frank Lloyd Wright (1867–1959), and Le Corbusier (1887–1965), among many others, particularly contributed to this formative process that culminated into ecological planning in North America. At the same time, this history, we conclude, provides a framework to understand and plan for current and future urban challenges by taking cognizance of the relationships between the built and natural environments.

12.2 The Three Initial Roots of Ecological Planning in North America

It has been speculated elsewhere that the emergence of ecological planning in North America was as a result of three main roots that composed its hidden and invisible past (Palazzo 1997, 2003)—to use an assertion of Thomas Kuhn (1962). This hidden and invisible past contributed unconsciously to the "uneven progress" (Zube 1986) of ecological planning. These roots—English landscape gardening that was transmuted into genuine American landscape architecture; the management of the vast public domain, particularly by some federal agencies; and the concept of regional planning as advocated by the Regional Planning Association of America and then later during the New Deal—represented a resultant effect of processes, practices, and notions that had taken place before the 1960s and 1970s in North America. In the sections that follow, we review these three roots.

12.2.1 The First Root: From English Landscape Gardening to American Landscape Architecture

The development phase of American landscape architecture took place between the end of the nineteenth century and the beginning of the twentieth century. The first exponents, of what was then considered American landscape architecture, considered aspects of natural landscape systems, how to protect them, and how to incorporate their features into the design of urban and country parks. Individuals who adopted this framework are recognized as being the founders of the profession of landscape architecture and as the ones who assembled its "toolbox" of scientific knowledge, historical references, and ethical and methodological principles.

Even in its earliest stages, set geographically on the East Coast and lasting for a period of about 100 years, landscape architecture was already shaping up as an authentically American profession and scientific discipline. It maintained a mark of originality during the course of the twentieth century becoming part of the ecological planning discipline and the more recently evolved field of environmental planning. American landscape architecture itself evolved from the English landscape gardening tradition (Newton 1971; Tobey 1973; Reed 1983; Jellicoe and Jellicoe 1987). It was then practiced, among others, by Sir William Kent (1684–1740), Alexander Pope (1688–1744), Charles Bridgeman (d. 1738), Henry Wise (1680–1738), Lancelot "Capability" Brown (1715–1783), and Sir Humphry Repton (1752–1818).

The new world's European heritage remained an integral part of the cultural patrimony. Nonetheless, it was mixed together with a uniquely American spirit that had as one of its most elementary features to be its relationship with the natural world. The United States inherited a lot of habits and traditions from England, which influenced and determined the development of the former colony. In addition to England's language, the United States kept its Common Law Judicial System, its institutional systems of territorial jurisdiction and division of powers (Platt 1991), and its system of private property valuation (derived mainly from Locke's *Two Treatises on Government*). These came into being at the end of the English colonial period and, over time, were amended and refined by its American appropriators.

In the field of landscape architecture, the American tradition owes much to England, at least in its preliminary stages. The English tradition was later elaborated and adapted to the natural and cultural traditions of the new world. The principal protagonists in the inauguration of the discipline got their inspiration from the landscape transformations realized in England during the eighteenth and the first part of the nineteenth centuries. Many of these principal protagonists traveled to England to get inspiration and knowledge from the creations of landscape gardens. Thomas Jefferson (1743–1826), Andrew Jackson Downing (1815–1852), Horace (H.W.S.) Cleveland (1814–1900), Frederick Law Olmsted (1822–1903), Calvert Vaux (1824–1895), Charles Eliot (1859–1897), and Warren Manning (1860–1938) were definitely influenced by them. Also important, and on a much larger scale, was the bond that Thomas Jefferson had with the European culture at a larger scale. This rapport figured prominently in his elaboration of the original elements of cultural inventiveness, which also evolved over the course of the next two centuries after the nineteenth century. This would later ripen to maturity specifically in the practice of American landscape architecture. This passage is well expressed in Krueckeberg's concise formula in the publication: "From the backyard garden to the whole USA" (Krueckeberg 1980).

12.2.2 The Second Root: The Management of the Public Domain

The public domain lands were distributed in many ways. Among them is the Homestead Act, which allowed a family to become the owner of 160 acres of land for free, on condition that the family built a house, settled the land, and demonstrated that it was cultivated for 5 years. This Act produced the most evident effects of all the federal regulations for allocating the public domain lands. It opened the West to considerable numbers of homesteaders. The opening of the frontier helped to create a pioneer culture and behavior that was biased and contradictorily on two foundations. The first was the appreciation of wilderness as values maker (Nash 1982). The second was based on a frontiersman's land ethic epitomized by the well-known attitude: "a man's got a right to do what he wants with his property" (Lewis 1973, 3).

Consequently, the Big Raid, which refers to the damaging impacts of these Acts and policies on the natural resources in the United States (Udall 1988), started with forests then with the exploitation of other resources, such as gold, oil, and natural gas. Agriculture and the cattle industry produced other serious damages, like soil erosion that generated the legendary Dust Bowl. Apparently, "Bemused by the Myth of Superabundance, Americans ignored the elementary laws of nature" (Udall 1988, 67).

In the last third of the nineteenth century, the most important phase of the great giveaway of land began. Some critiques emerged on what humans, in general, and Americans, specifically, were doing to nature. George Perkins Marsh (1801–1882) with *Man and Nature* (1864) and John Wesley Powell (1834–1902) with *The Report on the Arid Region* (1879) were at the forefront of this movement. They observed and documented the devastating effects of human activities on earth. This provided these protagonists the motivation and scientific grounds to advocate for the protection, defense, and wise management of natural resources. Furthermore, John Muir (1838–1914), Gifford Pinchot (1865–1946), Theodore Roosevelt (1858–1919), Aldo Leopold (1887–1948), and others started the Conservation Movement based on the rational use of land and resource. This subsequently strongly informed and facilitated the establishment of the primary federal land management agencies.

With the influence of these thinkers, the years between 1880 and 1930 in the United States represented the formative period of public policies in the field of natural resources protection, which were implemented by these federal land management agencies.

The evolution of techniques and ideas to manage forests, national parks, wilderness, and natural sanctuaries and to coordinate lands managed by the federal agencies became a remarkable set of experiences and practices for environmental and natural resource planners and scientists. This in turn influenced the formative processes of ecological planning in North America.

12.2.3 The Third Root: The Regionalism of RPAA

The Regional Planning Association of America (RPAA) was founded on April 18, 1923, at the Hudson Guild Farm in Mount Olive, New Jersey. The small group that inaugurated RPPA consisted of Lewis Mumford (1895–1990) who was the association's most prestigious spokesperson; Clarence Stein (1882–1975); the architects Henry Wright (1878–1936), Frederick L. Ackerman (1878–1950), and Robert

D. Kohn (1870–1953) from New York, John Bright (1869–1940) from Philadelphia, and E. Henry Klaber (1883–1971) from Chicago; the architect and planner Frederick Bigger (1881–1963) from Pittsburgh; the developer Alexander M. Bing (1879–1959) from New York; the economist Stuart Chase (1888–1985); Charles H. Whitaker (1872–1964) editor of the *Journal of the American Institute of Architects*; and the forester Benton MacKaye (1879–1975). Other members of the group were the housing experts Catherine Bauer Wurster (1905–1964) and Edith Elmer Wood (1871–1945).

During the brief life of the association (1923–1933), the RPAA members made important contributions to planning in North America and across the world. In Sunnyside Gardens at Long Island and Radburn, New Jersey, lessons from the English garden cities were employed and became the first examples of community planning in the United States. The *Report of the New York State Commission of Housing and Regional Planning* in 1926 was one of the first examples of state-level planning. Additionally, the *Northwest Plan* in 1920 and Benton MacKaye's *Appalachian Trail*, a model of interstate regional planning, were landmark outcomes from the RPAA (Ross 1983).

The faith in institution was another principle that inspired the members. Most of them worked with agencies or cooperated in federal or state programs. Their contributions can also be seen in the New Deal programs established by Franklin D. Roosevelt, from the Tennessee Valley Authority to the Civilian Conservation Corps, the Greenbelt New Towns program, the Rural Electrification Administration, and the National Resources Planning Board.

The ideas, theoretical efforts, and experiences of RPAA's members on regional planning are contributions that belong particularly to ecological planning heritage in North America. They demonstrate the relationship between the growing needs of a society and growing environmental constraints and showed the need for long-term visionary planning.

12.3 Conceptualizing Utopia's Root of Ecological Planning

The applications and literature of ecological planning embody an understanding of the quality and quantity of ecological and geographic resources; an understanding of how these resources are identified, formed, allocated, utilized, and/or misused; an evaluation of their efficacy for meeting different needs of society without compromising the availability of such resources for future uses; and creative and innovative ideas for achieving a balance between ecology, culture, and aesthetics (Zube 1986). Ecological planning is also iterative and futuristic, taking place throughout the entire planning and design process. It considers the potential factors that influence human and community interactions within and between the built and natural environments, while taking into account the implications for adaptability, equity, sustainability, and city or regional aspirations (Palazzo 2003). Lastly, ecological planning is political as different actors, values, interests, people's aspirations, and institutionally arrangements for decision-making including laws, regulations, policies, and agencies are inherently part of the planning and design process (Friedmann et al. 1989). Ecological planning thus presents a framework for understanding and responding to the dynamic and complex relationships between humans and their environment and helps planners envision how such relationships may manifest and perpetuate positive or negative effects.

Historically, such thinking has not been far from utopian philosophies and thinkings. From communitarian communities to grand visions of utopian philosophers, there is an underlining belief that humans and nature are part of an ecological system where relationships exist to either militate or enhance city functionality and human living. From this belief, holistic approaches to city problems have emerged, with utopian thinkers putting forward visions that incorporate the systemic interconnectedness between the built, the social and the natural environment. Three main points stand out in conceptualizing utopia's roots of ecological planning, namely: utopian visioning, city-country interconnectedness, and utopian holistic approaches.

12.3.1 Utopian Visioning

Cities have often been used as the laboratory for the imaginations of better futures. Such thinking recognizes that the built and natural environments as complex systems of competing relationships spanning the social, economic, physical, political, and environmental (Pinder 2005). Ecological planning which also envisions a better future, and to understand and resolve the "…interdependence of man and nature" (McHarg and Steiner 1998, 11), relates to utopian thinking in this way.

In the nineteenth and twentieth century, visions emerged to confront the challenges of urban disorder, social anarchy, and urban decay that followed the Industrial Revolution. Scientists such as Adna Ferrin Weber (1870–1968); sociologists and philanthropists including Charles Booth (1840–1916); political scientists and reformers such as Friedrich Engels (1820–1895), Matthew Arnold (1822–1888), Pierre-Joseph Proudhon (1809–1865), Charles Fourier (1772–1837), and John Ruskin (1819–1900); and writers such as Charles Dickens (1812–1870) and Victor Hugo (1802–1885) all presented ideas of a better future. The initial discourse on confronting urban challenges resulting from the Industrial Revolution began in Europe. This process started 50 years later in North America (Choay 1965).

From the first known utopian community in North America through the 1960s and 1970s, the challenge of growing urban decay influenced utopian ideals of bringing humans closer to nature (Hayden 1976). Urban pollution, water quality issues, natural disasters, and the overall decay of the urban physical environment inspired new urban visions that relied on building a strong relationship between humans and their environment.

European migrants from England, France, Germany, Scandinavia, and Eastern Europe were pioneers of utopian thinking in North America. These immigrants saw North America as an opportune land to realize new visions of promoting harmony with nature. Often discussed in terms of communitarian utopian communities, these thinkers "proposed a complete restructuring of city and country in response to the environmental problems created by the industrial revolution" (Hayden 1976, 9). Examples in North America go as far back as when John Davenport and Theophilus Eaton founded New Haven in 1638, after arriving from England. This is one example of several communitarian utopian communities that swept through North America in the early seventeenth century to the early part of the nineteenth century.

The utopian visions of new communities by European migrants provide a profound understanding of conceptions of the ideal city. The varying architectural, spatial, economic, social, and cultural perspectives of their visions ultimately endorsed a city model that draws humans closer to nature. The plans for such cities were rigorous and detailed, showing how urban life could persist—devoid of the crowded and chaotic cities and slums (Fishman 1982). As Robert Fishman points out, these ideal cities "were convenient and attractive intellectual tools that enabled each planner to bring together his many innovations in design, and to show them as part of a coherent whole, a total redefinition of the idea of the city" (1982, 6). For instance, the Shakers' way of living "aimed at nothing less than transforming the earth into heaven" (Hayden 1976, 67).

New visions of cities were also proposed by Robert Owen (1771–1858), Charles Fourier (1772–1837), Benjamin Ward Richardson (1828–1896), Etienne Cabet (1788–1856), and Pierre-Joseph Proudhon (1809–1865) who, using scientific techniques and rational thinking, defined a fundamental set of the basic necessities of individuals and cities, while promoting green spaces as necessary to healthy living (Choay 1965). This view was also applied in regional planning, where many began to advocate for exploring environmental factors before undertaking any form of planning. This tradition was espoused by Patrick Geddes (1854–1932) and later Lewis Mumford and Benton MacKaye, who emphasized that both humans and nature are critical for giving the city a new life and "an alternative urban morphology" (McHarg and Steiner 1998, 6). William Morris (1834–1896), John Ruskin (1819–1900), and Ebenezer Howard (1850–1928) emphasized the role of culture in maintaining the natural resource connection to cities and their landscape and morphology and saw cities as organic units that needed to be cultivated and managed during the changes due to the Industrial Revolution. This was to be achieved by promoting harmony between environmental resources available to town and country, as well as effective and efficient land management, regeneration, and planning (Choay 1965, Tod and Wheeler 1978). Frank Lloyd Wright (1867–1959) also advocated for a proposal that promoted rural and natural landscapes in cities, criticizing their structure and conditions that emerged as a result of the Industrial Revolution (Wright 1928).

These utopians conceptualized a transformation of cities through proto-ecological principles and also placed importance on the political systems that impact city life. In fact, when Thomas More first coined the term Utopia (no place) in 1516, he depicted an imaginary society of different political dispensations that focuses on mitigating the social problems of present-day society (More 1989). Utopians further advocated that design, architecture, and planning must also relate to the political dimensions and culture of urban life if harmony were to prevail between the built and natural environments (Pinder 2005).

These new visions imbibed in the luxury and harmony of nature informed, in part, the formative processes of ecological planning in North America. The desire for a new city devoid of decay, a city that promotes harmony between the built and the natural environment as well as harmony between different environmental units, became paramount to the theory and practice of ecological planning. Indeed, "visions of the ideal life have guided the modern planning movement" (Neuman 1991, 344) and have remained core to the evolution of ecological planning. As past writers have discussed, visioning can be an "imaginative projection of a new place or state" (Pinder 2005, 15), a desire for "good society" (Kumar 1991, 27), a desire of "an alternative urban morphology" (McHarg and Steiner 1998, 6), or a vision for a better way of life (Bloch 2000). In this sense, urban planners and designers, landscape architects, and politicians are constantly visioning—as they design and consider alternative better living environments.

In this way, "*we are all utopians*, as soon as we wish for something different and stop playing the part of the faithful performer or watchdog" (Lefevre 2000, 75). Thus, the separation of utopia from visioning becomes arbitrary (Rowe 1976). Evidently, utopia provides a lens to appreciate how these visions of cities and the environment have been romanticized over the years. Ecological planning, in a similar vein, which was and is still driven by visions of a different and better futures for both people and cities, is simply a reflection of utopian visioning principle.

12.3.2 City-Country Interconnectedness

Another aspect of utopia that influenced the formative processes of ecological planning was the need to build dynamic connections between cities (i.e., the built environment) and the country (i.e., the natural environment). Many utopias emphasized a need to marry the town and country (Howard 1902) as a way of managing the environmental challenges that confront cities. In the late nineteenth century, the desire for better cities was in response to the urban challenges of crowded and fast-growing cities characterized by slums, air pollution, and general spatial disorder which had taken humans far away from nature. Since the first notion of utopia by Thomas More, who imagined a city comprised of six thousand families integrated with the country (Pinder 2005), these urban challenges have influenced utopian thinkers to idealize the attractiveness of the country.

A study of utopian communities in America also reveals that the dynamic connection with the natural environment manifested in town and regional planning. This informed efforts to modify and plan for the development of residential, agricultural, and industrial uses in cities (Hayden 1976). Visions like Ebenezer Howard's Garden City, Frank Lloyd Wright's Broadacre City, and Le Corbusier's Radiant City were also inspired by this principle. This town-country principle has guided "the modern town planning movement and alter[ed] its objectives" (Mumford 1965, 29), making "environment" an objective of planning and design in North America.

Utopian thinkers considered how such resources such as fresh air, water, and greenery can be made part of a city's landscape. As in utopian thinking, ecological planners looked at how these same natural elements could be employed in green spaces or formal gardens and parks to replace urban decay and make nature the lens through which urban spatial transformation was viewed, interpreted, and manifested. To mitigate urban degeneration, therefore, was to appreciate cities as part of a human ecology or as being ecological (MacKaye 1940). Ian McHarg's famous book *Design with Nature*, first published in 1969, and Philips H. Lewis Jr.'s book *Tomorrow by Design: A Regional Design Process for Sustainability* (1996) both provided a methodological and a theoretical framework for integrating the country into the city—bringing key elements of utopia into mainstream ecological planning.

Efforts to integrate the country into the city became especially apparent in the 1960s and 1970s when the United States was faced with the challenge of environmental crisis. Americans became aggrieved with issues of polluted environments from pesticides, oil spills, and radioactive overrun into their environments making it dangerous for human living (Egan 2007). Public outcry and street demonstrations spurred reforms to mainstream the natural environment into human activities. Such reorientation extended to planning and design during the environmental crisis and became crucial as "new professional skills were needed to provide various services in connection with environmental assessment, waste disposal planning, air and water quality management, and so on" (Marsh 2005, 3). The need to respond to the environmental crisis led to the consolidation of three different professional fields, namely, environmental (or ecological) planning, environmental science, and environmental engineering and technology (Marsh 2005).

12.3.3 Utopian Holistic Approaches

A third aspect of utopian thinking that is relevant for the formative processes of ecological planning relates to the approach to problem identification and planning response. Utopian thinkers identified cities as systems that function effectively when there is a balance among social, economic, political, spatial, environmental, institutional, and cultural units. Accordingly, harmony among these units was the basis for confronting pollution, slums, and blights in the built environment. These thinkers also promoted social transformation by ensuring equitable access to economic and productive resources. This was one of the foundations for promoting harmony between town and country. Their visions therefore represented "a total environment in which man [or humans] would live in peace with his fellow man [or human] and in harmony with nature" (Fishman 1982, 7). These thinkers also promoted harmony through holistic approaches that are now applied in ecological planning.

Through holistic approaches, utopian thinkers identified and examined the different units of cities to ensure effective resource management. For instance, Frank Lloyd Wright and Le Corbusier saw the important role technology plays in effective resource management leading them to conclude that the advent of the Machine Age offered technological opportunities to enhance cities and their landscapes rather than their despoil (Mumford 1925, Crawford 1970). Lewis Mumford shared similar observations of the role of technological revolution in the "fourth migration" (Sussman 1976). Fourth migration referred to the kind of urban life that emerged from using automobiles and electrical energy which "made possible for a new decentralized urban form of satellite communities" (Bauman and Muller 2006). These thinkers saw technology as necessary for managing issues of social justice and urban degeneration through reconciling humans, nature, and machine (Fishman 1982). Such reconciliation was also a way to envision how cities could promote industrialization while maintaining a natural environment that supported relaxation and/or agriculture in cities. This reconciliation thus reflected cooperativeness, integration, and interaction between the city and country. Such utopian ideals demonstrate to planners and designers that nature cannot be excluded from the built environment when aiming to build a better and healthy society. The need for reconciliation between the opportunities of the country and the city and understanding their resource potentials are also principles strongly embraced in ecological planning.

The notion of envisioning ideal cities, holistic appreciation of environmental units, and the city-country interconnectedness are part of the utopian patrimony. These three aspects of utopia discussed so far underpinned the principles, approaches, and philosophies of the what, how, and why of ecological planning in North America.

Specifically, utopian communities between 1790 and 1975 viewed their cities in different ways: as gardens with a commitment to the wise use of land, as a machine with potential for innovation, and as an ideal home that committed residence to collectivity (Hayden 1976). The Garden City by Ebenezer Howard is one of several utopian concepts that have persisted through time. While the entirety of the Garden City vision as well as other utopian visions may have not manifested, their principles and philosophies have remained relevant to ecological planning and to planning and design principles and practice in general.

Examples of the Garden City idea implemented in the United States include Reston in Virginia and Columbia in Maryland—both of which were built after World War II (Ward 1992). These were informed by environmental concepts of greenbelts and green spaces, as envisioned by Howard. In addition, Howard's principles are associated with the "Greenbelt cities" constructed in the United States in the 1930s under the America Resettlement Administration's Greenbelt Town Program (Myhra 1974). Open space planning for parkways and park systems were also informed by ecological reasoning (Elson 1986).

For ecological planning, utopian thinking remains a dominant part of its intellectual patrimony. The utopian thinking that informed the ecological planning discourse on city and country interconnectedness has translated into cities where the built environment is interspersed with parks, urban forests, and gardens. This has allowed for efficiency and beauty to persist together. The adoption of urban agriculture and farms, as well as suburban and edge city development, also provides evidence of utopian ideas being incorporated into planning, particularly into ecological planning. Moreover, ideas such as green intentional communities, which can be considered as utopias (Sargisson 2007), are evidence of the ecological principles of utopian thinkers.

12.4 Conclusion

In this paper, we have argued that utopian thinking significantly contributed to the formative process of ecological planning in North America and can therefore be seen as its fourth root (since the other three have been extensively discussed in Palazzo 1997). This discourse on utopia and ecological planning demonstrates the importance of appreciating past processes that ultimately culminated into an existing body of knowledge and practice. The evidence presented reinforces the fact that ecological planning was not entirely a product of the 1960s and 1970s, but rather an outcome of different bodies of thought from the past. Varying perspectives about this history can be found in texts such as Ndubisi's *Ecological Planning* (2002), Ian McHarg and Frederick Steiner's *To Heal the Earth* (1998), and Ndubisi's *The Ecological Design and Planning Reader* (2014), where these authors allude to how ecological (or environmental) planning was inspired by the philosophers and professionals mentioned in this paper. The discussions in this paper broaden the perspective on ecological planning and help to uncover "hidden" principles, philosophies, and methods that can be made relevant today. Hence, this can enrich the whole body of theory and practice associated with ecological planning.

In summary, different utopias provided and proposed different designs for urban spatial and social morphology, for urban restructuring and transformation, and for the organization of cities. Nonetheless, they all share a common convergence, the objective of harmony between the built and natural environments. Furthermore, utopian thinking provides a framework for appreciating how visions of ideal cities have transmuted into ecological planning principles, the importance of understanding effective resource management for a better way of living, and a need to appreciate holistic approaches to city problem management. This framework is similar to ecological planning principles as they have been developed and applied in the past 50 years. We conclude on the note that the values of the past—in this case, utopian values—present greater insight for managing today's urban challenges and provides another framework within which to respond to future urban challenges. It is from this understanding that utopian thinking is proposed as the fourth root of ecological planning, as it provides for theoretical, philosophical, and practical elements in the present-day field of planning, especially in ecological planning.

References

Bauman JF, Muller EK (2006) Before renaissance: planning in Pittsburgh, 1889–1943. University of Pittsburgh Press, Pittsburgh

Belknap RK, Furtado JG (1967) Three approaches to environmental resource analysis. Harvard University Landscape Architecture Research Office, Conservation Foundation, Washington, DC

Bloch E (2000) The spirit of Utopia. Stanford University Press, Stanford

Choay F (1965) L'Urbanisme. Utopies et realites: Une anthologie. Editions du Seuil, Paris

Crawford A (1970) Ten letters from Frank Lloyd Wright to Charles Robert Ashbee. J Arch Hist. 13:64–76. doi:10.2307/1568314

Eckbo G (1950) Landscape for living. Duell, Sloan and Pearce, New York

Egan M (2007) Barry Commoner and the science of survival: the remaking of American environmentalism. MIT Press, Cambridge, MA

Elson MJ (1986) Green belts: conflict mediation in the urban fringe. Heinemann, London

Fishman R (1982) Urban Utopias in the twentieth century. MIT Press, Cambridge, MA

Friedmann J, Chinitz B, Blanco H, Neuman M (1989) Planning, politics, and the environment. J Amer Plan Asso. 55(3):334–338

Hayden D (1976) Seven American Utopias: the architecture of communitarian socialism, 1790–1975. MIT Press, Cambridge, MA

Howard E (1902) Garden cities of to-morrow: being the second edition of "To-morrow: a peaceful path to real reform. S. Sonnenschein Limited, London

Jellicoe GA, Jellicoe S (1987) The landscape of man: shaping the environment from prehistory to the present day. Thames and Hudson, New York

Krueckeberg DA (1980) From the backyard garden to the whole U.S.A. A conversation with Charles W. Eliot II. J Amer Plan Asso. 46(4):440–448. doi:10.1080/01944368008977076

Kuhn TS (1962) The structure of scientific revolutions. University of Chicago Press, Chicago

Kumar K (1991) Utopianism. Open University Press, Milton Keynes

Lefebvre H (2000) Everyday life in the modern world. Bloomsbury Publishing, London

Lewis PF (1973) The geographer as landscape critic. In: Lewis FP, Lowenthal D, Tuan YF (eds) Visual blight in America, commission on college geography resources, Paper No. 23. Association of American Geographers, Washington, DC, pp 1–8

Lewis PH (1996) Tomorrow by design: a regional design process for sustainability. Wiley, New York

MacKaye B (1940) Regional planning and ecology. Ecological Monographs 10(3):349–353

Marsh WM (2005) Landscape planning: environmental applications, 4th edn. Addison-Wesley Publishing Company, Hoboken

McHarg IL (1969) Design with nature. Natural History Press, Garden City

McHarg IL, Steiner FR (1998) To heal the earth: the selected writings of Ian L Mcharg. Island Press, Washington, D.C

More T (1989) Utopia. In: Logan GM, Adams RM (ed) Cambridge University Press, Garden City. (First printed in Latin in 1516, and in English in 1551)

Mumford L (1925) The fourth migration. Survey Graphic 7:130–133

Mumford L (1965) The garden city idea and modern planning. In: Howard E, Osborn FJ (eds) Garden cities of To-morrow. MIT Press, Cambridge, MA, pp 29–40

Myhra D (1974) Rexford Guy Tugwell: initiator of America's Greenbelt New Towns, 1935 to 1936. J Amer Inst Plan 40(3):176–188. doi:10.1080/01944367408977467

Nash RF (1982) Wilderness and the American mind, 3rd edn. Yale University Press, New Haven/London

Ndubisi F (2002) Ecological planning: a historical and comparative synthesis. The Johns Hopkins University Press, Baltimore

Ndubisi F (ed) (2014) The ecological design and planning reader. Island Press, Washington, DC

Neuman M (1991) Utopia, dystopia, diaspora. Amer Plan Asso. 57(3):344–347. doi:10.1080/01944369108975505

Newton NT (1971) Design on the land: the development of landscape architecture. Belknap Press of Harvard University Press, Cambridge, MA

Palazzo D (1997) Sulle spalle di Giganti. Le matrici della pianificazione ambientale negli Stati Uniti. Franco Angeli, Milano

Palazzo D (2003) The hidden past. A thesis about the roots of environmental planning in North America. Paper presented at the ACSP-AESOP 3rd Joint Congress, Leuven (Belgium), 8–12 July 2003

Pinder D (2005) Visions of the city: utopianism, power, and politics in twentieth-century urbanism. Routledge, New York

Platt RH (1991) Land use control: geography, law, and public policy. Prentice Hall, Englewood Cliffs

Reed M (1983) The Georgian triumph, 1700–1830. Routledge & Kegan Paul, Boston

Ross JR (1983) Benton MacKaye: The Appalachian Trail. In: Krueckeberg DA (ed) The American planner. Biographies and recollections. Methuen, New York/London, pp 196–207

Rowe C (1976) The mathematics of the ideal villa, and other essays. MIT Press, Cambridge, MA

Sargisson L (2007) Imperfect Utopias. In: Leonard L (ed) Utopias, ecotopias and green communities: exploring the activism, settlements and living patterns of green idealists, advances in ecopolitics, vol 1E. Emerald Group Publishing Limited, Bingley, pp 1–25

Sussman C (1976) Planning the fourth migration: the neglected vision of the regional planning association of America. MIT Press, Cambridge, MA

Tobey GB (1973) A history of landscape architecture: the relationship of people to environment. American Elsevier Publication Company, New York

Tod I, Wheeler M (1978) Utopia. Harmony Books, New York

Udall SL (1988) The quiet crisis and the next generation. Peregrine Smith Books, Salt Lake City

United States Department of Interior and Bureau of Land Management (2013) 2012 public land statistics. Bureau of land management. http://www.blm.gov/public_land_statistics/pls12/pls2012.pdf. Accessed 18 Jan 2016

Walker P, Simo LM (1994) Invisible gardens: the search for modernism in the American landscape. MIT Press, Cambridge, MA

Ward SV (1992) The garden city: past, present, and future. E & FN Spon, New York

Wright FL (1928) In the cause of architecture. The logic of the plan. Architectural Record 43(1):49–57

Zube EH (1986) Landscape planning education in America: retrospect and prospect. J Land Urb Plan. 13:367–378. doi:10.1016/0169-2046(86)90054-X

Chapter 13
Integrating Methods: Aligning the Effects of Scale and Bias

Philip D. Plowright

Abstract This chapter examines the cognitive structure of design methods as an entry point to understand how disciplinary bias, issues of scale, and complex situations can be negotiated. Moving from the foundational structure of divergent and convergent thinking, the discussion is focused on the underlying frameworks and location of judgment criteria in design processes. These are the key to understand how diverse methods can engage each other across disciplinary bias and scales to align design outcomes. Ultimately, scaling methods is about the ability to transfer values between missions and goals to strategies and tactics within a design method.

Keywords Scale • Bias • Design

13.1 Introduction (What Is Good Design and Why We Should Not Worry About It)

We can start with the statement that good design is important. We also have to understand that is a pretty useless statement. The definition of good design has been debated for centuries with no resolution. We can return to the French Academy, who formalized this discussion in their first meeting in 1671 (Kruft 1994) addressing *bon goût*, or good taste, which was behind design decisions. At that point, François Blondel led the debate, and a year later, there wasn't any agreement on the idea. As far as the Academy got at this point was to determine that good taste, and therefore good design, was that which pleased intelligent people. The next question, of course, is who was considered intelligent? Judgment was linked to the political and social structure of the time – those who were intelligent were those who held power and authority. Over 60 years later, the question was still being debated in the Academy.

P.D. Plowright (✉)
College of Architecture and Design, Lawrence Technological University,
21000 West Ten Mile Road, 48075 Southfield, MI, USA
e-mail: pplowrigh@ltu.edu

R. Ingaramo, A. Voghera (eds.), *Topics and Methods for Urban and Landscape Design*, Urban and Landscape Perspectives 19,
DOI 10.1007/978-3-319-51535-9_13

In 1734, the Academy published their definitions of four fundamental concepts behind the theory of architecture, which included good taste. Finally, we had a formal definition which stated that good taste was the harmony between parts and was constructed from physical proportions, formal distribution of the design and sensitivity to past uses (Kruft 1994). This definition did not extend past attempting to code beauty through visual interpretation to Platonic ideals (Plato, *Philebus* 64e) and code the use of precedents as a source for design knowledge. Now, I would like to believe that we have developed stronger structures of social justice since the seventeenth century, ones that allow more perspectives to be reflected through our design values. I would hazard to say, however, that we have not gotten much further in understanding how values are integrated into methods. In 2006, we still find generic language that identifies desired qualities in good design "as being scale sensitive, strategic, politically sensitive, stakeholder-driven, adaptable, specific-value-driven, systemic, frugal, and flexible" (Hellmund and Smith 2006). I don't believe there will be much disagreement that these factors are desirable and do address developments in social justice. But what does it mean to have a frugal and sensitive design *in context*? The terms are so vague and disconnected from any design method to be wishful thinking, like saying that one wishes to be happy but without investigating what experiences, objects, and interactions lead to happiness.

One of the issues is that over the past centuries, and growing in strength through the twentieth century, design methods in architecture in Western society have been seen as personal, subjective and unique to individuals. This attitude is encapsulated in the Kantian belief of genius as the source for good design. There is a large gap between general belief statements of values and the processes that it takes to achieve those values. There is also a difference between an appropriate design method for desired results and good decision-making within a design method. Considering that many believe design methods are not knowable at an individual level, we can make things even more complicated by introducing projects with multiple scales, diverse disciplines, and complex stakeholder representation. What happens in terms of methods when these are involved within a single project?

This is an interesting question. We can begin to explore both the issue of design process structure and complex project delivery by considering the basic structure of a design method from a cognitive point of view. We will find that all design uses the same foundational thinking structures (Plowright 2014). However, applied methods involve multiple iterations of that thinking structure. Methods that engage more than one person or a complex set of requirements need mechanisms that structure basic design thinking with larger priorities and values, including the mapping on non-disciplinary content into disciplinary values. This is an issue of alignment of values and a strong testing mechanism. There are, however, serious human, disciplinary, and information structural factors at play that strive to disrupt *any* applied method. We have to remember that any obscurely defined goal – such as any project focused on sustainable development – will make understanding methods more complicated. Without a common agreement and the ability to measure success of outcomes, it is very difficult to create projects where everyone can agree that they are "scale sensitive, strategic, politically sensitive, stakeholder-driven, adaptable, specific-value-driven, systemic, frugal, and flexible," even if we have no idea what those terms mean. Sustainability, for example, has multiple definitions and interpretations depending on the discipline using the term with little shared agreement of operational factors between involved parties – even in closely aligned disciplines such as those involved in the built environment (Adhya et al. 2010; Plowright and Adhya 2012).

Before we start looking at multiple scales, get more people involved, and engage public policy and finances as well as political opinions and personal desires, we can consider foundational cognitive structure and then move from thinking sequences to more complicated structures. The first question is what is design and how does it work? Design is not exclusive to our professions with the word in their title; it is a basic human capacity (Ward et al. 1999). Design is about simulating a future state and testing the possibility of that state against known factors in the world. As such, it is a decision-making structure and is based on the thinking processes that all decision-making uses: exploration, analysis, projection, and simulation. In fact, the act of designing might be considered the thing that makes humans human (Buchanan 1992), and all decision-making could be seen as an aspect of design. The question is one of focus and application. To put it another way, as explained on the children's program *Sesame Street*, "Design means think about and plan to put something together in a certain way before you build it" (Chris, Sesame Street-O-Saurus 2013). This is true, but also a limited definition as outcomes do not always need to be predetermined. I will return to this comment once we have the foundational thinking structure illustrated.

So if this is design, what is a design method? A method is a collection of cognitive exploratory and analytic tools, set in a sequence, aligned to a value set, and supporting certain outcomes. When it comes down to basics, design is about two things. First, on a structural level, design is a process of decision-making. The famous École des Beaux-Arts *parti*, that which guides the architectural proposal, most likely came from the French expression *prendre le parti de*, meaning "to assume a position between several choices." The origin is based on the designer making a choice or a decision by which to guide the development of the building proposal. Within every design process, thousands of choices are made and the role of the method is to align those decisions into a greater whole to make a relevant and significant proposal. Second, on an operational level, design is an act of empathy that aligns with human and ecological values – the emotional, physical, and sociocultural. It is impossible to design well if the designer cannot project beyond themselves into the needs of a larger community. Sustainability *should* introduce a much broader concept of empathy that move beyond human-to-human welfare to consider a diversity of flora and fauna populations as well as large earth systems such as hydrology. A well-developed method allows coherence in decision-making so the outcomes are aligned with the initial values as well as developing a shared point of discourse by contributors and stakeholders.

13.2 The Anatomy of Design Thinking

What is the most basic operation in design thinking? Design uses two types of cognitive patterns known as divergent and convergent thinking (Fig. 13.1). The coupling of these types of thinking is the basic mechanism for all design work. Since the 1970s, divergence and convergence have been well documented and can be found in studies of creativity and innovation as well as design (Jones 1973; Rowe 1987; Finke et al. 1992; Dorst and Dijkhuis 1995; Dorst and Cross 2001; Lawson 2006). Divergent thinking is exploratory. It is involved when people use brainstorming, questioning, superimposition, or other techniques to generate a series of options or information around an idea. Divergence is nonjudgmental and based on generating as many ideas as possible, even those that seem outlandish or unexpected. Convergent thinking is the opposite – meant to narrow choices and to make a selection. It generally occurs after divergent thinking as it uses the cloud of ideas generated by the exploration in order to reduce that content to a choice. At the end of a convergent process is a decision or selection. Actually making the selection requires some more structure – a goal, a bias, an association with other selection elements within the design, and so on – but the basic operation of design is the deployment of these two thinking patterns.

We can return to the Sesame Street scenario, *Sesame Street-O-Saurus*, to see the thinking structures at work. To give a little context, Rosita, Elmo, and Grover are pretending to be paleontologists and have uncovered a collection of "fossils," which are really just discarded everyday objects – a ladle, a xylophone, keys, a spring, and wood spoons. They decide to make a model of a dinosaur, so Grover just jams everything together and ends up with a mess. At this point, Chris (the "adult") suggests that Grover uses *design* to plan how the model will be created from the parts they have. So the three friends map out in their heads what the final model will look like before building it. Rosita and Elmo are quick to pick up on how to move beyond functional fixativeness and can see the ladle as the dinosaur head and xylophone as the body. Grover is a little slower. However, the slowness is used to allow Grover to

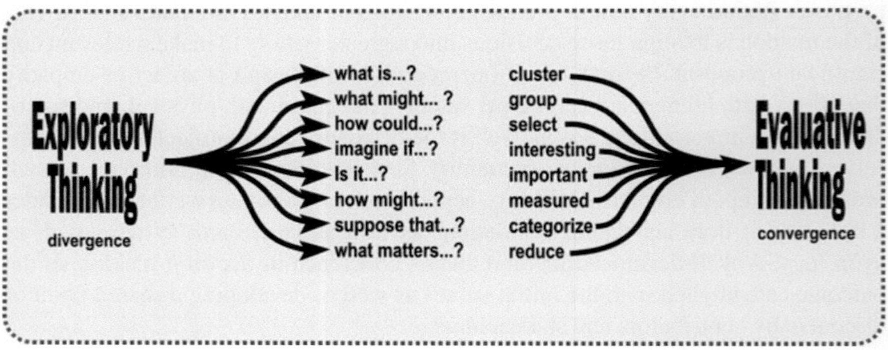

Fig. 13.1 Divergent and convergent thinking patterns (Source: Author's elaboration)

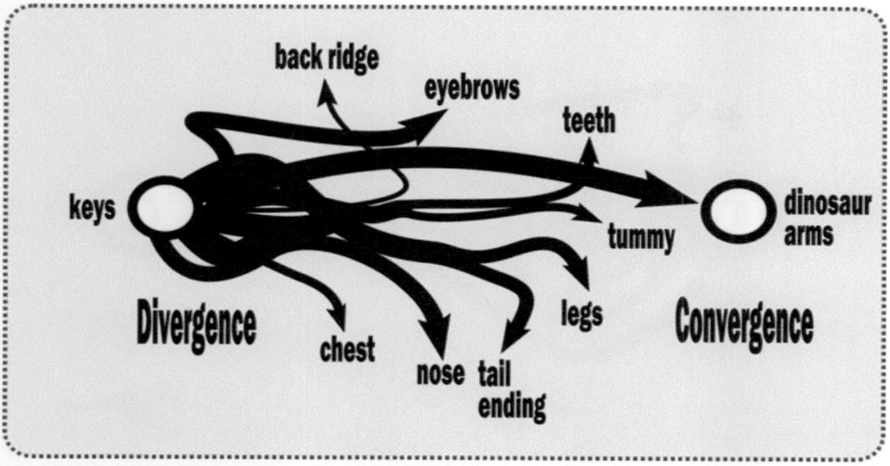

Fig. 13.2 Divergent-convergent mapping for the keys as dinosaur parts (Source: Author's elaboration)

illustrate divergent and convergent thinking engaged to make decisions while maximizing possibilities and innovation (it is an instructional program for children, after all). Grover doesn't quite get the idea of how to match shapes of objects to body parts of the dinosaur. Divergence occurs when he considers each of the shapes and runs through all the possible locations of where those shapes might be used. The explicit example is the two keys and where they might map to some part of the dinosaur body. Grover runs through the options – the tummy (doesn't need a pair), the eyebrows (not one of the most distinguishing features of a large reptile) – and then selects the best mapping, the keys as arms. This is what is called an attribute mapping as the teeth of the keys evoke the shape of hands of a Tyrannosaurus. For our purposes, it illustrates the divergence of thinking – considering all the places that it might be possible for the keys to be used and then the convergence of thinking, isolating, and selecting one choice with the strongest possibilities while discarding the others (Fig. 13.2).

The divergence-convergence coupling is a very foundational cognitive structure in design (Fig. 13.3). The basic shape is an expansion of information, gathering as many options around an idea as possible, and then an evaluation of those options to make a selection. There are many conceptual tools that can be used in the process, and many are interchangeable or developed by individual designers. However, all divergent-based tools are used to develop alternatives, while all convergent-based tools refine and reduce content to a selection.

Some scalar and disciplinary bias can be found in the particular tool selection and their deployment as part of a method. It is the selection of a particular tool which creates boundaries for what information is addressed. Any tool, regardless to whether it is a physical object like a *hammer* or a conceptual process like *sectioning*, has a context in which it is best used and to which outcomes are limited. The

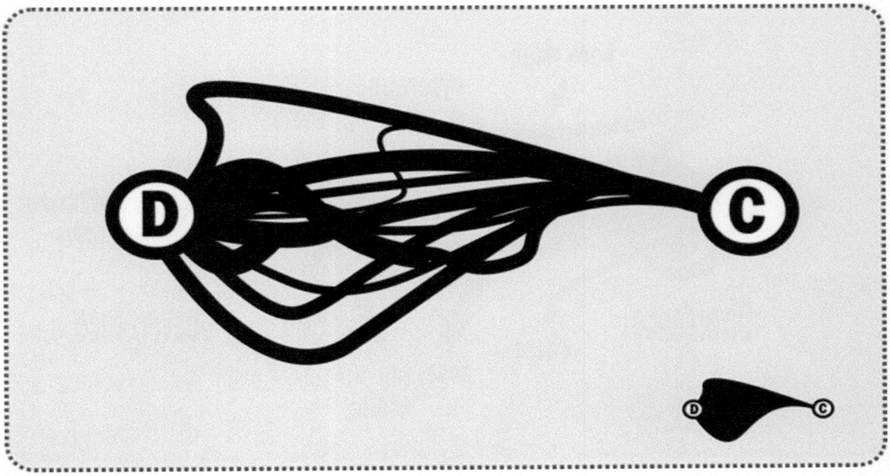

Fig. 13.3 Basic divergent-convergent structure (Source: Author's elaboration)

A: Welcome to Earth
Q: What is this place?
A: Earth or this building?
Q: Right now, where am I?
A: You are in the public library.
Q: Is every building a Library?
A: No, there are lots of kinds of buildings. A library is just one of them.
Q: What makes this a library?
A: There are books here.
Q: What is a book?
A: It is printed material, bound, and read.
Q: You are here and you are not a book.
A: I came here to meet my book group.
Q: Is that a stack of books?
A: No. It is a group of people that get together and discuss a book. Its theories, validity, the story and message it talks about.
Q: The book talks?

Questioning: Alien as the Other

A: Kind of. As you read it your mind is submerged in the world it creates. You have an internal dialogue about its message.
Q: Is this a bad world? Should I get out of here? It sounds like no one wants to be here.
A: No. It's not a bad world. It is nice to imagine something new and fun that is different than your ever day life.
Q: Soo... all these people here are trying to escape from their lives?
Some are. Some are here to learn or exchange ideas, like me. Some are here because it gives them a place that is quiet. There are some here that don't have access to the information in these books or the internet any other way.
Q:Everyone is here because they want something that they can't get anywhere else. Nowhere else on Earth have these things?
A: You can find things like this other places. Discussions at lectures and schools, information on-line, quiet in your home. Here they find all of those things and can decide what they want to do, what is interesting to them for free.
Q: What does interesting mean?
A: Interesting... attractive, motivating, exciting, out of the ordinary. People need diversions.
Q: Free, you mean they don't pay anything to be here? With so much offered why don't they make money on stuff?
A: Not everyone has money to pay for these things but as a society we have decided information and communication is important. It doesn't matter if you agree with all of it or even if it is a subject you want to read. I hate romance novels but some people love them and it is important that they have access to them.
Q: So everything ever written is here?
A: No. They can't get everything in here. Some things are too rare. Some aren't deemed "acceptable". Most libraries around here will try to get what you need through library exchange and find things on-line.

Fig. 13.4 Divergent tool – questioning (Source: Erin Smith)

expectation when using a hammer is to penetrate wood with a nail to adhere two or more elements together. We might be able to perform the same function with a wrench or shoe heel, but it wouldn't be efficient or natural to do so. We also wouldn't expect an informational return that involved screws or nonmechanical adherence – things which have been created for, and associated with, tools such as screwdrivers and glue. As an analogy, the same principle of tools use works in cognitive processes as in physical.

For example, we can use either graphic or textual approaches to divergence, but immediately we can see that text and image hold different types of information. The classic questioning technique, which stresses a form of reduction to reveal unexpected knowledge about a known situation (Fig. 13.4), can capture abstract principles and express feelings. The *dérive* mapping tool (Fig. 13.5) allows for the graphic exploration of a territory. The unexpected alignment of landscape and built environment elements would be impossible to explore in the same way through text.

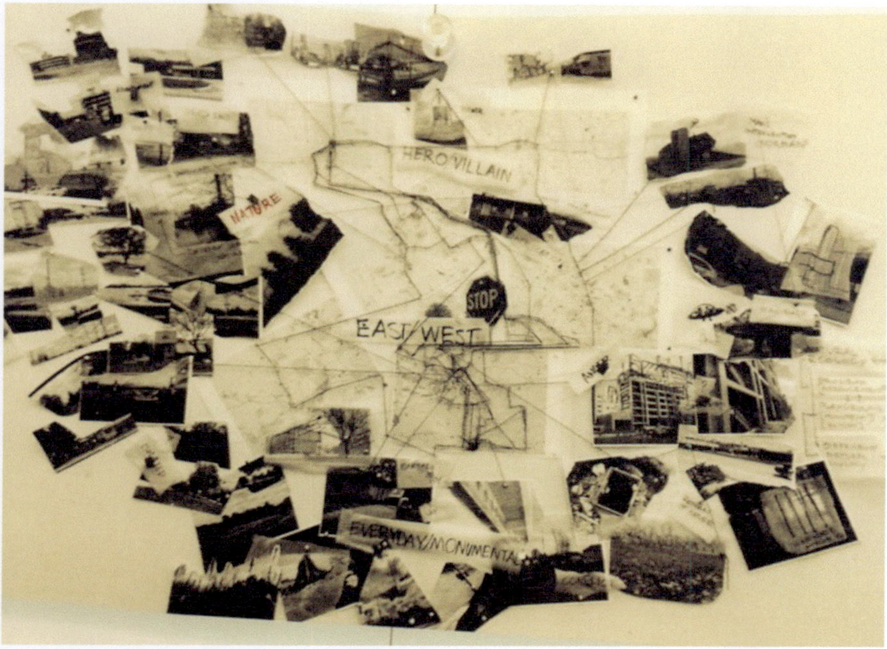

Fig. 13.5 Divergent tool – situationist *dérive* mapping (Source: Irsida Bejo et al.)

The versioning divergent tool (Fig. 13.6) allows discovery of a multitude of variations of a single element stressing formal manifestation. Each of these operates in context with a particular scale and a particular set of information. However, they are all divergence based.

It could be said that it is coupled thinking styles which makes design be design. Recent research has shown that in so-called irrational processes, steps that are used to displace the expected are still based on divergent and convergent cognitive processes and can still be grouped into large categories (Jovanovic Weiss and Plowright 2014; Plowright 2015). The strength in the divergent-convergent process is its access to innovation. The diversity of thinking that is inherent to the relationship between thinking styles has been shown to increase quality and innovation of response (Stempfle and Badke-Schaub 2002). The stronger the process of exploration and analysis, the greater the potential for a design outcome to be both novel and significant. However, the cognitive richness – different ways of thinking – is offset by issues of communication and values. People have a tendency to value what they know and align themselves with others who hold similar values. This is the issue working between disciplines – we can often get social cohesion, which is good if we want to like each other, but struggle to get task cohesion, which is needed for quality of outcome.

The divergent-convergent action that occurs hundreds of times within any design process is non-scalar, free of bias, and context-independent *until* applied. Any

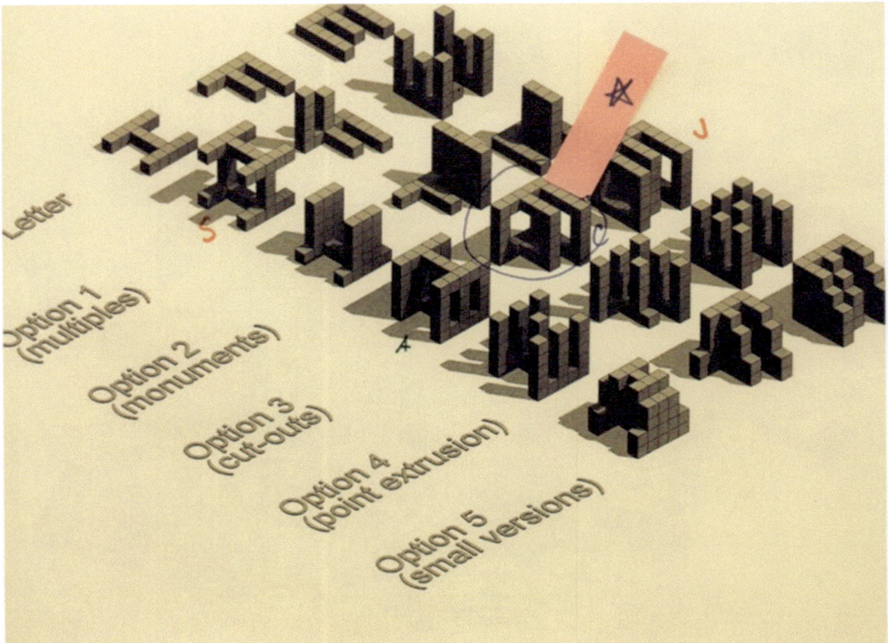

Fig. 13.6 Divergent tool – versioning (Source: Amy Swift et al.)

divergent tool can replace another divergent tool, while any convergent tool can replace another convergent tool – it all depends on what the designer wishes to achieve and what type of information is being addressed. The reason why one aspect of the divergent cloud is chosen over another has to do with larger-scale content in the design process, as well as external factors such as belief systems, disciplinary bias, cultural priorities, and social attitudes.

13.3 Source Information and the Seeds of Methods

While we might consider design methods to be unique to an individual, they are not. All architectural design methods are structured on a collection of divergent-convergent-coupled thinking styles organized by the type and source of information chosen to focus the design process, creating frameworks (Plowright 2014). The frameworks are meta-approaches, while the thinking styles are the basis of applied tools.

There are three major approaches or frameworks created by information sources, each one having a historical development as well as a different relationship to the judgment criteria – that part of the process that allows for decisions to be made. This moves us to the next dimension of a method structure, connecting several

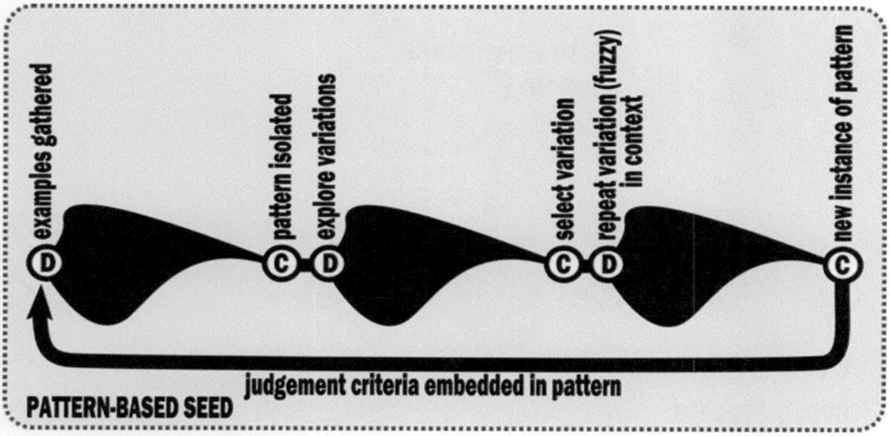

Fig. 13.7 Pattern-based basic cognitive seed (Source: Author's elaboration)

divergent-convergent couples together in persistent frameworks, varying scale and complexity. Major orientational approaches to information sources for design prioritize either studying current human-to-form, human-to-environment, and environment-to-form interactions (forces), from past and best-practices examples of interactions (patterns) or from some overarching projected desire (concept). These approaches were formalized during the late Enlightenment and the Rationalist movement of the nineteenth century. While some aspects of these structures can be found earlier in architectural treatises, it isn't until a Western society fully invested in scientific method and formalized pedagogy that these frameworks are founded as clear, repeatable structures.

The first truly rational architectural design method was produced by Jean-Nicolas-Louis Durand (1760–1834) and published in his book, *Précis of the Lectures on Architecture* (1802–1805). Durand's method stressed information coming from existing examples of successful projects and best practices, what we call precedent studies (Durand 2000). He presented architectural design as the application of *patterns* and *rule sets* – this architectural framework uses composition and arrangement of elements in space to drive design decisions. Patterns, as a value, are seen by the designer to hold the best type of information to allow for relevant outcomes for the final design, and those patterns create compositional rules. A framework focusing on pattern application then limits the tool selection as well as structures how decisions would be made and what type of information would be available to the designer – stressing composition and typology over cultural and social use of space (Fig. 13.7). Cultural and social content is still present; only it is held in the patterns rather than being applied independently as part of the design process (Plowright 2014, 39–40). An explicit judgment criterion is absent because all values come from preexisting examples and are held within the pattern, making testing not necessary. The pattern has proven to be successful, and if the basic principles are

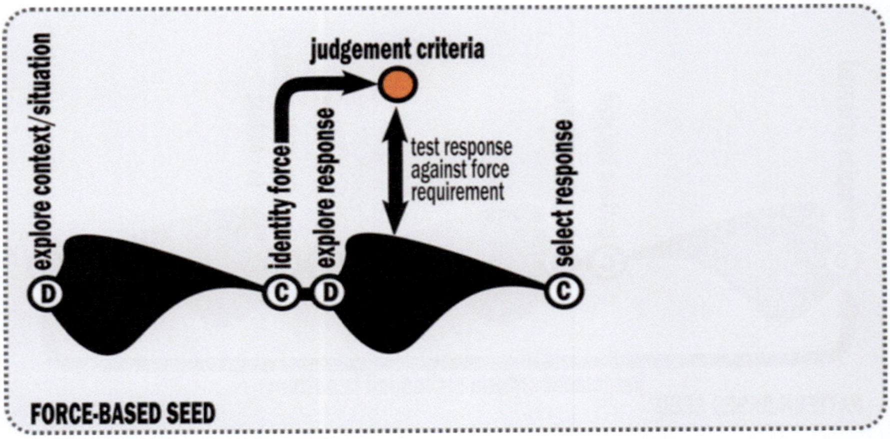

Fig. 13.8 Force-based basic cognitive seed (Source: Author's elaboration)

repeated, then any variation of the pattern should be successful again. Decision-making is held to the choice of pattern and the association between pattern variations.

The second major framework for architectural design is one where the major information source is from forces (Fig. 13.8). Forces are any nonformal information that allows formal response – the common ones are environmental, such as sunlight, air movement, and microclimate, or social such as circulation, gathering, view, openness, and human activities. Although this approach differs conceptually from the application of pattern-based rules, it developed from the same rational, scientific approach as the one detailed by Durand (2000). The force-based framework focuses on systems thinking and often conceptualizes forces as pressures, assets, constraints, and flows. The point of the framework is to make those forces accessible and ordered so a designer can act upon them through formal response. Documentation of the framework can be found in Eugène-Emmanuel Viollet-le-Duc's book, *Histoire d'une maison* (1874), although the attitude toward forces can also be found in writings as far back as the Renaissance architect Leon Battista Alberti (Plowright 2014). On a process level, a force approach considers architectural form as the direct manifestation of forces, flows, or pressures (Sullivan 1918; Alexander 1964; Groak 1990). Identifying these pressures through the introduction of a series of *constraints* and *assets* allows decisions to be negotiated, moving toward a final proposal. Many designers who approach their work using a *force-based* methods tend to believe that design is a problem-solving process and that design is simply the resolution of conflicting forces (Plowright 2014). Judgment criteria operate through the identification of the force, allowing the response to maximize the positive outcome (asset) while suppressing the negative (constraint). It is the choice and analysis of the forces being chosen which then sets the parameters of success.

The final framework, concept, is one that has aspects found in the theoretical notion of character forwarded by Gabriel-Germain Boffrand (1667–1754) and analogy as found in the Renaissance architects Alberti, Francesco Di Giorgio

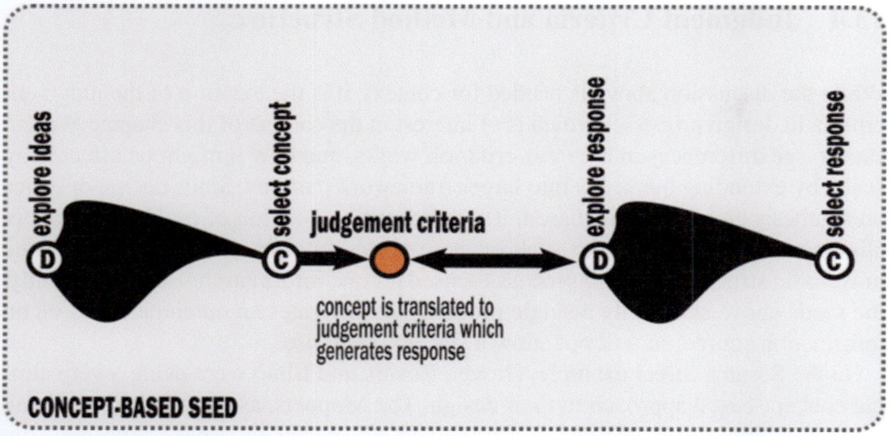

Fig. 13.9 Concept-based basic cognitive seed (Source: Author's elaboration)

(1439–1502), Filarete (1400–1469), and Palladio (1508–1580) as well as the Revolutionary architects Étienne-Louis Boullée (1728–1799) and Claude-Nicolas Ledoux (1736–1806). Both positions stress coherence as the goal of the outcomes, and the major source of the model for architecture was poetry and literary arts. The concept of character became institutionalized into architectural education through the École des Beaux-Arts of the early 1800s through the use of the *parti, or "what characterizes a building"* (Cret 1941). We would no longer consider the parti as a contemporary approach to concept as it stressed formal composition over conceptual ideas and emotive effects. The contemporary framework of concept, instead, owes its structure to a rational investigation in literature through an essay titled "The Philosophy of Composition" (1846) by the American poet, author, and critic Edgar Allan Poe. In this essay, Poe reinforced the philosophical position Alexander Gottlieb Baumgarten had developed over a century before (Baumgarten 1954). Picking up on *artistic unity or lucid order* as a value in design outcomes, Poe called for the development of a *unity of impression*. In order to have unity, there would need to be a focus. For Poe – as for Boffrand, Blondel, Boullée, and Ledoux – that focus or central idea would be first and foremost on *effect*. In architecture, this meant that concept is a top-down approach which uses a "big idea" to structure decisions within the design process – unlike patterns and forces which are emergent, bottom-up processes. The *concept-based* framework revolves, then, around the creation of a central idea which is used to organize the parts of design proposal. All aspects of the design are then judged against, and should reinforce, the central idea (Fig. 13.9).

The diagrams above show only the most basic, one-dimensional action from a design process focused on their information source. They are not yet a framework but are what would be called *seeds*, the pieces that hold the nascent structure for the larger pattern. It is from each of these that a framework develops.

13.4 Judgment Criteria and Method Structure

While the discussion above is needed for context, it is the location of the judgment criteria in design processes which is of interest in the context of this chapter. We can start to see differences in how this criterion works, and how it might be affected by scale, by extending our seeds into larger framework patterns. Since design operates on synthesis and is systems-based, it requires the association of multiple factors in alignment or interaction with each other to create a proposal. A framework is the most basic structure of a full process focused on one information source. Currently, the seeds above show only a single process that generates an outcome based on an information source, so will not allow a systems response.

In the Sesame Street example, Grover, Rosita, and Elmo were using a very simple concept-based approach to their design. The Muppets, as part of their role-play being paleontologists, decided to use their found objects to make a dinosaur – an idea that is coherent with their overall theme, a hint that concept would govern the approach. The desired outcome, the *concept*, was simple – the form of a dinosaur. While we were not privy to the divergent discussion that arrived at this decision (the left half of Fig. 13.9), we can assume a form of this process occurred – i.e., Elmo asks Rosita what they should do next; all the Muppets think about it, exploring and discarding options; and Grover says "let's make dinosaur!" The other two abandon their ideas and agree, arriving at "select concept." By stating that the object of the design was to make this particular shape, they had set an overall judgment criterion by which all aspects of the design processes were tested. The right side of Fig. 13.9 describes the general process that occurs to map against the judgment criteria. Each object is explored to how it supports the whole, with as many options as possible explored to and then have the best selected (i.e., strongly meets the needs of the judgment criteria) and the other options abandoned. This is repeated for all aspects of the project until there is a point of substantial resolution. Figure 13.10 shows what the concept structure looks like with three lines of enquiry – say, the keys, the ladle, and the xylophone.

This is now a framework showing the full relationship between the parts. What is interesting is how in the concept approach the judgment criteria are centrally located to control the choices in the entire process. There is only one (the orange circle marked "j.c." on the diagrams), and all divergent-convergent processes return to it to test choices for satisfying the design criteria and to make decisions.

We can look at three lines of inquiry for the pattern and force approaches as well in order to understand how different the cognitive structure of these design frameworks are compared to concept. While the concept framework has explicit and well-developed judgment criteria, the pattern-based framework (Fig. 13.11) has implicit judgment criteria, embedded in the choices of types or patterns to use to drive the process.

This means that there are no active guiding criteria once the choice of pattern has been decided. All pattern-based approaches start with the same information – the reduction of phenomena to a pattern which generates at rule. The rule is then the

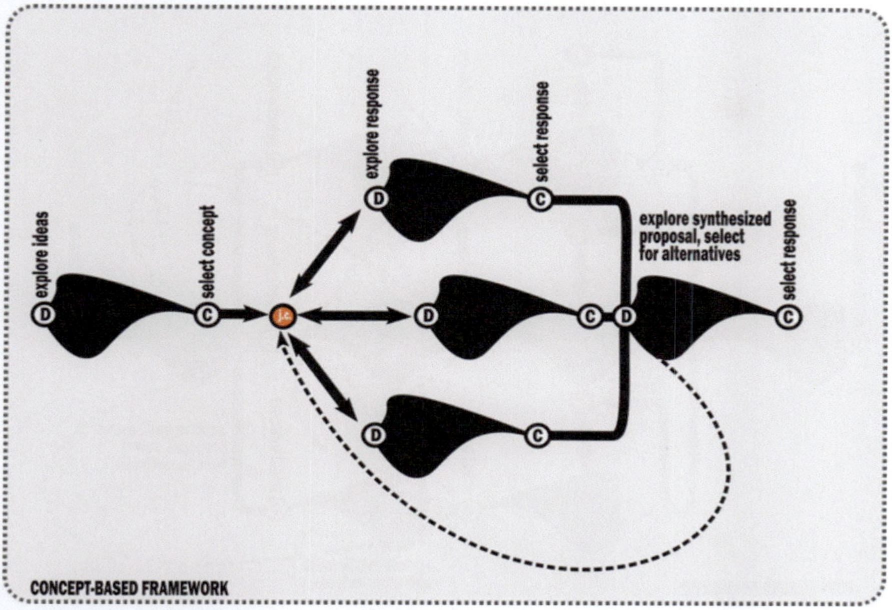

Fig. 13.10 Concept-based design framework using seeds in three lines of inquiry (dimensions) (Source: Author's elaboration)

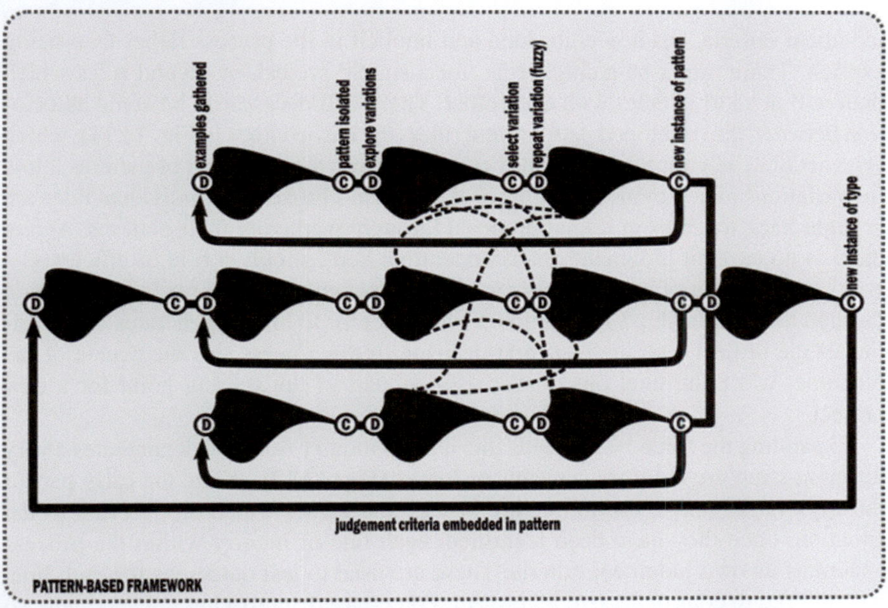

Fig. 13.11 Pattern-based design framework using seeds in three rules (dimensions) (Source: Author's elaboration)

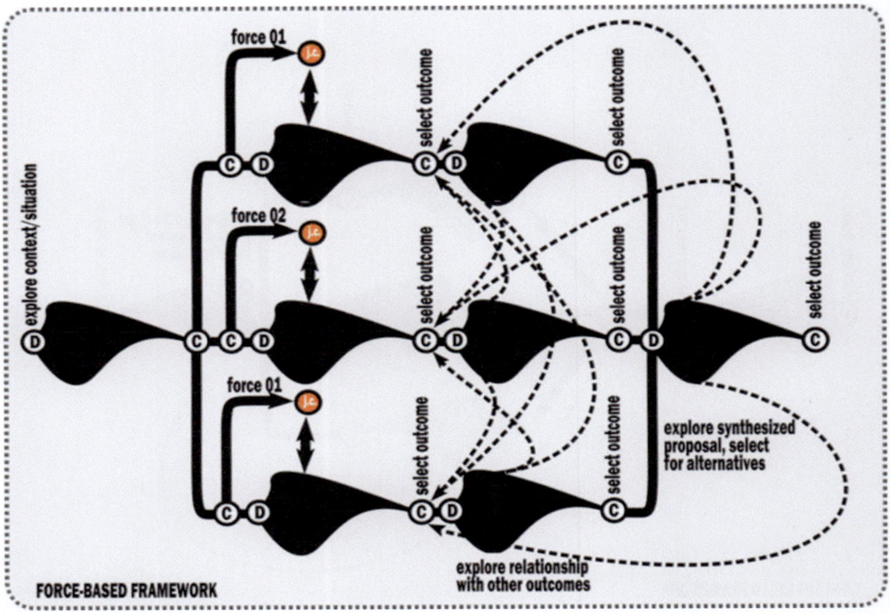

Fig. 13.12 Force-based design framework using seeds in three lines of inquiry (dimensions) (Source: Author's elaboration)

judgment criteria, but it is embedded and implicit in the process rather than being explicit. There might be a single rule, for a simple project, or several rules which then will need to interact with each other. There will necessarily be some association between the developed patterns and rules (the dashed lines in Fig. 13.11), which is an act of synthesis. Since the rules generated from the reduction of patterns allow for variations in deployment, as long as the rule isn't broken, the individual lines are brought back together in a final proposal through overlaying their patterns. Again, there is no explicit judgment criteria operating here, simply a rule of efficiency – how can an aspect of one rule support the deployment of another rule so both gain through the relationship? The choice of what patterns to look at generates the priorities of the project – so the very first decision in the process sets the course of the outcome, while the final outcome is used as part of the starting point for a new project.

Expanding the force-based seeds into a three-inquiry framework generates a very different structure to either concepts or forces (Fig. 13.12). While the early part of the exploration and decision-making is selecting forces which are relevant to the situation, once they have been identified, each line of inquiry within the process generates its own judgment criteria. These are used to test outcomes for each line, but not used to control the overall project. Force-based approaches are emergent, so

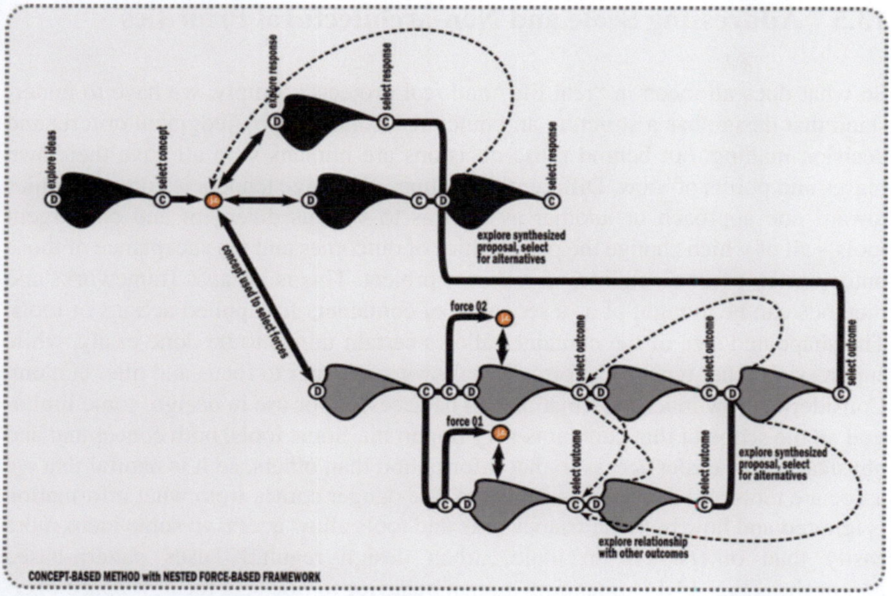

Fig. 13.13 Concept-based method with nested force-based framework. First judgment criteria sets subcriteria boundaries (Source: Author's elaboration)

the final proposal is the result of the interaction of the elements within the process rather than being controlled by a larger, overarching idea like in concept. As in pattern, though, the various sub-outcomes are negotiated with each other to stress synthesis through efficiency (one element performing two or more tasks).

All the frameworks, regardless to source seed, are valid processes – one is not better than the other; they simply prioritize different approaches that align to different types of informational sources. In addition, any divergent or convergent tool can be used in the sequences – again depending on what the designer wishes to achieve. The frameworks all end at the same point – a choice and an outcome. This is independent of scale. Frameworks can be used as a method by simply adding lines of inquiry, as many designers don't vary very far from the basic structure. But more complex cognitive process structures can be developed from the basic frameworks, and experienced designers can nest several frameworks together – using a concept structure to set overall goals and force or pattern structures to look at particular aspects of the design (Fig. 13.13). It is possible to connect all three of the framework styles as *the outcomes are independent to both the process and scale* until they are employed in a situation. In this case, the major judgment criteria will set the boundaries for the nested frameworks, already limiting their focus and setting boundaries for their own judgment criteria.

13.5 Addressing Scale and Non-architectural Priorities

So what does all mean in "real life" and real projects? Simply, we have to understand that design has a structure and outcomes controlled by judgment criteria and decision-making, but behind those decisions are humans who all have their own biases and points of view. Different disciplines also have tendencies or sympathies toward one approach or another as well as to various divergent and convergent tools – all of which change the possibilities of outcomes and the acceptance of those outcomes by other disciplines in a shared project. This is because frameworks and methods can be thought of as a sequence of containers for applied actions or tools. The shape and size of the container allows certain things to be done easily, while suppressing other types of information allows a designer to focus and filter content. Considering how much information can be accessed for use in design, some limitation on the scope of that content is very important. Some tools, both conceptual and physical, allow easier access to that information than others, so it is natural that we associate tools and frameworks together. The danger comes from what information is ignored and how both the frameworks and tools allow access to some ideas more easily than others. For example, urban design regularly uses pattern-based approaches (Fig. 13.14), mapping human behavior to rule sets for how space operates socially, but this is rarely found in architectural design today. In fact, there is a cultural bias against this type of information within architecture as the framework stresses preexisting and historical information rather than the development of unique situation. This doesn't mean it isn't valid as an approach, just that the discipline does not value it for ideological reasons.

All disciplines have a naturalized way of thinking about their core content and easier access to some information than others. This is the reason why dealing with variable scalar content, multiple disciplines, and broad goal in a single project introduces a series of communication and acceptance obstacles that must be overcome.

These factors introduce noise, nonoperational concepts, seemingly diverse priorities, and apparent conflicts between general information found in the world and

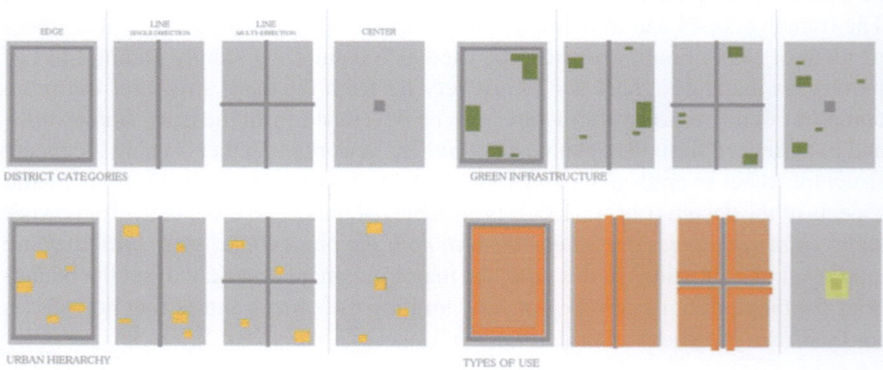

Fig. 13.14 Urban analysis of urban patterns of district development (Source: Joy Sportel)

narrow applications within disciplines. This makes it difficult to ensure that what is accomplished at one scale (e.g., the design of a discrete building) aligns with values at another scale (such as management and preservation of a watershed). It becomes harder to decide what matters when the priorities set by the project are not native to the various disciplines that are engaged. Applied methods are context specific – disciplines have developed ways of doing things that work for the type of outcomes that discipline values. However, most explicit methods are invisible to those disciplines, and all applied methods contain bias which needs to be negotiated by overarching conceptual structures. Often there is a mismatch between common tools that *apply* ideas and concepts – such as the policy, plan, specification, or schedule – and the larger idea. This is because moving on the gradient between the biosphere and the human hand (which might include continent, political region, locality, neighborhood, building scales) changes the type and form of relevant information. It also changes the *explicit* testing content. In order to maintain coherence, involved disciplines need to translate the priorities which have particular syntax, terminology, priorities, and value structures. This brings us to the requirement for a strong project method so to guide design method interaction.

13.6 Territorial Strategies

Disciplines, as well as organizational entities, define territories which imply bias and held values. The territory stresses that responsibilities as well as the type of information that is considered is limited. Outcomes and focuses from naturally occurring disciplines will be the result of historically developed tools and production processes. Information that isn't naturally occurring within the boundary of the discipline will be either invisible, considered irrelevant, or misinterpreted unless some form of translation occurs (Fig. 13.15). Otherwise, the internal biases and

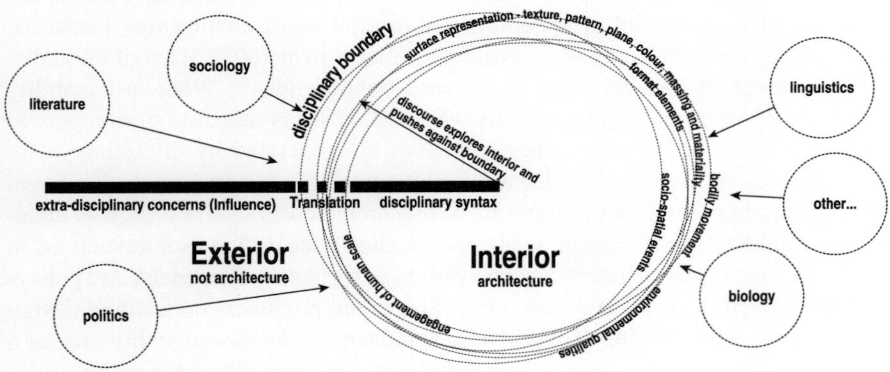

Fig. 13.15 Mapping territorial domains of disciplines (architecture) (Source: Author's elaboration)

values will tend to structure outcomes on areas considered comfortable for this territory – social groups will trend toward social decisions, while technical groups toward scientific facts. In terms of methods, it is important to know both the scale of the major intentions and how it factors into normative disciplinary processes. It isn't good to have expectations for an outcome only to not be able to achieve it because the tools being used do not support the type of information desired. If the disciplinary tools do not have direct access to the type of desired content, then the method needs to be changed or the tools adjusted. This is a tactical issue at the divergent-convergent level as well as strategical issue at the project framing level.

While most disciplines have inherent access to their core values and methods, it is a different issue when working between disciplines. In this case, large-scale value structures need to be determined before methods are engaged and then translated into each discipline's natural syntax to ensure a level of relevance and alignment of expected outcomes. If a disciplinary process is struggling with scale and value effects not found naturally in their processes, an intervention must be made to insert those values by either changing the tools they use in the divergent-convergent process or by translating overall goals into discipline friendly terms. The key to understanding the effect of scale on methods is to consider the interface of discrete design methods within overall project methods. Often these terms are used interchangeably; however, they refer to very different scales of processes. While it is admirable to have a process that attempts to take in all players, one has to remember that an embedded design method inherently has biases that support disciplinary vision over any nondisciplinary value. In inter- and transdisciplinary projects that are known to span scales and disciplinary priorities, an orientation tool at the mission level is critical for alignment and maintaining larger-scale sociocultural and ecological values.

For example, studying sustainable development and urbanism revealed that each of the four major players in the construction of the built environment had latent values built into their methods, rhetoric, syntax and tools (Adhya et al. 2010; Plowright and Adhya 2012, 2014a). The systems-based context of sustainability often necessitates understanding and addressing the built environment as an amalgamation of ideas shared by specific allied disciplines of architecture, landscape architecture, urban planning, and civil engineering (Inam 2002). Each of these disciplines brings its own priorities to the sustainability debate. What sustainability means to engineering is not necessarily what sustainability means to planning, or to landscape, or to architecture. These priorities are generated by disciplinary language, values, and tools embedded as latent ideological positions in a design-based practice and in generation of proposals. While architects prioritize design as three-dimensional formal orientation, landscape architects see design as interventions in open space networks; planners conceive design as regulatory framework and policy-making reflecting social and economic values; and engineers understand design as efficiency in production. This eclectic approach toward sustainability creates a partitioned model with conflicts, contradictions, and radically different priorities depending on who defines the nature of sustainable development (Fig. 13.16).

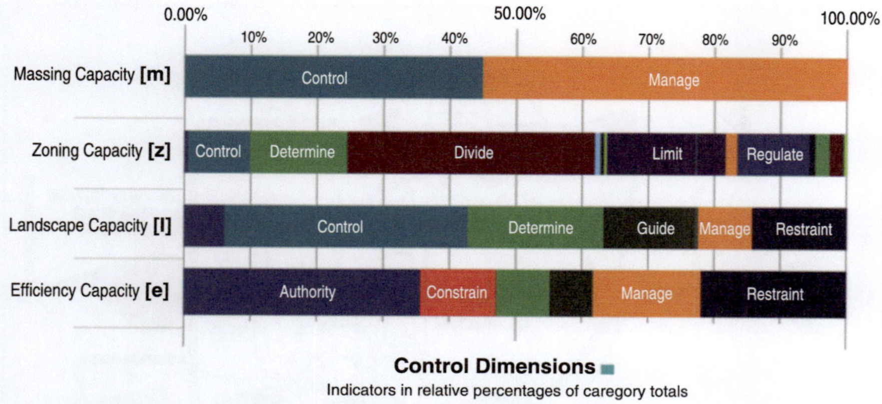

Fig. 13.16 Semantic indicators for control dimension across disciplines involved in sustainable development (Source: Author's elaboration)

This means that before any design method is developed, there is always a philosophical position or framing that isolates what matters as a successful outcome and set of values to be engaged through the process (Chong and Druckman 2007). There are many terms that can be used to describe this: philosophical belief, starting bias, belief structure, or community values. It is usually translated into a mission statement or statement of purpose which is left general using disciplinary neutral language. The interesting thing, and what is important for inter- and transdisciplinary processes, is this is where the major objectives and values are held. These are not *within* the method itself but found before individual design starts (Plowright 2014). The mission statement is also the main tool for scaling priorities from large cultural-environmental statements to applied values within a method regardless to the discipline involved.

This process is not unlike a nested design method such as the one illustrated in Fig. 13.13, except on a much larger scale. In fact, it could be argued that all methods, from the small-scale framework to the large-scale project structure, are fractal, infinitely repeating structures of the same basic processes – divergent-convergent couples with hierarchical judgment criteria affected by basic informational approach at each level of occurrence.

The mission statement, then, acts as a judgment criterion at the project level and the central point of testing by which all subprocesses are structured. It limits the starting position of all the other design methods involved but does not limit the information approaches nor the outcomes, *as long as they align with the values encapsulated by the mission statement*. A set of values or belief structure for significance allows an internal process – be it an architectural, urban, planning, hydraulic engineering, political policy-making, agricultural development, or environmental engineering – to align with shared external values (Fig. 13.17). It is also the point of transfer for discussions within a single project with multidisciplinary teams with divergent terminology and internal values. The importance of this action cannot be

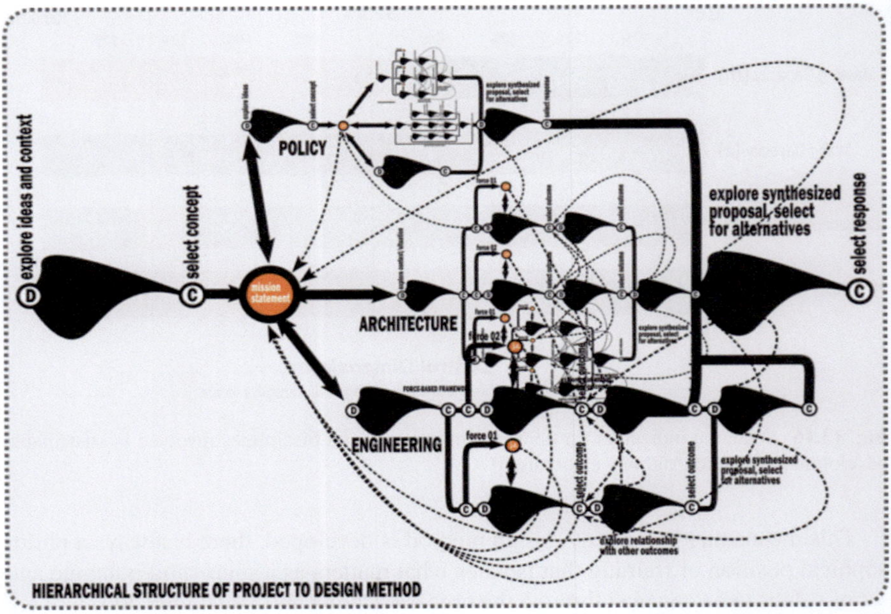

Fig. 13.17 Project method with nested design methods involving multiple disciplines (Source: Author's elaboration)

overstated. An agreed, large-scale, general mission statement will allow parallel disciplines a common point of communication and value agreement. It also allows a disciplinary design strategy to be tested constantly against the overall goal, allowing multiple discrete design-based outcomes to parallel and support each other.

A clear example of the use of a mission statement as part of a metastructure is a goal-based meta-tool such as the Action Agreement (Ingaramo and Voghera 2012a) or Action Contract (Ingaramo and Voghera 2012b). This tool identifies values which transcend disciplinary boundaries and allows the alignment of priorities that act as a value marker. The Action Agreement sets up the ability to translate values into disciplinary positions and is the critical communication tool when dealing across disciplines. It is an orientation procedure that maintains the testing standard, allowing a process of coherence to occur at multiple scales and within disciplinary bias.

While the Action Agreement is a particular event within an applied method, its need is reflected in the cognitive structure of all design methods as detailed above. All methods operate through judgment criteria and decision-making, whether this is explicit or implicit. When crossing disciplinary boundaries and operating at multiple scales, a clear value marker is critical. Discussions that will take place at this scale should be expansive, using generic language. The more disciplines involved in the approach, such as sustainable development, the more necessary is a reduction to foundational and shared values (Plowright and Adhya 2012, 2014b). For example, sustainability should not be considered in disciplinary terms first, but in human and ecological ones. A shared mission statement would capture the key points that would span the understanding of multiple stakeholders and agents. Only then should the priority be translated into disciplinary access and method be engaged.

13.7 Conclusion

For some reason, methods in the applied design fields are not addressed with the equal clarity as the same cognitive processes found in other disciplines, such as business and military theory (Friedman 2003). Yet all these processes share a common goal of "design" – the structured pursuit of a projected future outcome.

The key to understanding both design methods as a discrete event and project methods as a collection of design methods is to understand judgment criteria, how information sources structure the relationship between divergent-convergent couples and how information transfers across disciplinary boundaries. The first move in any project which allows alignment with desired outcomes is a large-scale judgment statement (the *mission* or *vision* statement if using business, military, or political disciplinary language). In design, we would call it a philosophical position, starting bias, or a belief structure. This is a concept that identifies what matters and is a very general, large-scale position. It should be able to span disciplines and be stated in shallow or non-disciplinary specific language. The next part would be to develop the *goals* or *objectives*. There are often several goals to support a single mission and are guiding principles to focus outcomes. These also are part of project method identification and span disciplinary boundaries. The values from these project positions are needed to frame the testing structure that is used to limit what decisions should be made and why in the thinking structures occurring below this level.

It is the mission and goal levels that are the sites for interdisciplinary discussion and agreement. Strategy and tactics are engaged within disciplines in order to achieve those goals in alignment with other disciplines – but they all use the same cognitive processes to achieve their outcomes. Scaling methods are about transferring values between missions and goals to strategies and tactics that occur within a design method. Disciplinary bias is then aligned with overall values to ensure a larger degree of coherence between expected outcomes and disciplinary delivery.

We introduce methods based on a design framework affected by framing philosophies in order to limit our range of choices and to align those choices with values that have social and ecological relevance. Methods immediately narrow options and starting position, creating a series of limits rather than being open to any influence. This is important in a complex system – and the world we live and work in is a complex system. It is important to understand that without a guiding agreement that spans disciplinary values, project choice will gravitate toward those more naturally found in the historic values of the operating discipline, be it architecture, biology, agriculture, or politics.

References

Adhya A, Plowright P, Stevens J (2010) Defining sustainable urbanism: towards a responsive urban design. In: Proceedings of the conference on sustainability and the built environment. King Saud University, Saudi Arabia, pp 17–38

Alexander C (1964) Notes on the synthesis of form. Harvard University Press, Cambridge, MA

Baumgarten AG (1954) Meditationes Philosophicae de Nonnullis ad Poema Pertinentibus. English edition: Baumgarten A G (trans: Aschenbrenner K, Holther WB). University of California Press, Berkeley/Los Angeles

Buchanan R (1992) Wicked problems in design thinking. Des Issues 8(2):5–21

Chong D, Druckman JN (2007) Farming theory. Annu Rev Polit Sci 10(1):104

Cret PP (1941) The Ecole des Beaux-Arts and architectural education. J Am Soc Archit Hist 1(2):14

Dorst K, Cross N (2001) Creativity in the design process: co-evolution of problem–solution. Des Stud 22(5):425–437

Dorst K, Dijkhuis J (1995) Comparing paradigms for describing design activity. Des Stud 16(2):261–274

Durand J-N L (2000) Partie Graphique des Cours d'Architecture Faits a l'Ecole Royale Polytechnique Depuis sa Réorganisation; English: Précis of the lectures on architecture; with Graphic portion of the lectures on architecture. Texts & Documents, Getty Research Institute, Los Angeles

Finke RA, Ward TB, Smith SM (1992) Creative cognition: theory, research, and applications. The MIT Press, Cambridge, MA

Friedman K (2003) Theory construction in design research: criteria: approaches, and methods. Des Stud 24(6):507–522

Groak S (1990) The idea of building: thought and action in the design and production of buildings. E & FN Spon (Taylor & Francis), London

Hellmund P, Smith DS (2006) Designing greenways: sustainable landscapes for nature and people. Island Press, Washington, DC

Inam A (2002) Meaningful urban design: teleological/catalytic/relevant. J Urban Des 7(2):36

Ingaramo R, Voghera A (2012a) Planning and architecture. Searching for an approach. Alinea International, Florence

Ingaramo R, Voghera A (2012b) Finding a design method: "Integrated multiscale project". In: Proceedings of the EAAE/ARCC international conference on architectural research, Politecnico di Milano, Italy, pp 526–529

Jones JC (1973) Design methods: seeds of human futures. Wiley-Interscience, London

Jovanovic Weiss S, Plowright P (eds) (2014) Inhabiting everyday monuments. CRITPraX. http://issuu.com/critprax/docs/critprax14_book_ebook-issuu

Kruft H-W (1994) A history of architectural theory. Princeton Architectural Press, New York

Lawson B (2006) How designers think, the design process demystified. Architectural Press, Oxford

Plato, Hackforth R (1972) Plato's Philebus; Uniform Title: Philebus. Cambridge University Press, London

Plowright P (2014) Revealing architectural design: methods, frameworks & tools. Routledge, London

Plowright P (2015) Structuring the irrational: tactics in methods. In: Proceedings of the 3rd international conference for design education researchers, vol 1. Aalto University, DRS, Cumulus, DESIGN-ED, Chicago, pp 397–415

Plowright P, Adhya A (2012) Setting priorities: sustainability, environmental health, and embedded value judgments for the urban design process. In: Proceedings of the EAAE/ARCC international conference on architectural research, Politecnico di Milano, Italy, pp 118–121

Plowright P, Adhya A (2014a) Sustainable architecture as ecology: defining shared systemic priorities of public health and social ethics in places. In: Khan A Z, Allacker K (eds) Architecture and sustainability: critical perspectives. ACCO, Leuven (Belgium), Den Haag (Netherlands)

Plowright P, Adhya A (2014b) Wilderness and its absence: dominance of anthropocentric human-nature relationship in sustainable development. In: Proceedings of EDRA45NewOrleans, Environmental Design Research Association (EDRA), New Orleans

Poe EA (1846) The philosophy of composition. Graham's Magazine 28(4):163–167

Rowe P (1987) Design thinking. The MIT Press, Cambridge, MA

Sesame Street (2013) Sesame Street-O-Saurus. Season 43 (2012–2013). February 4, 2013 http://www.sesamestreet.org/media/video_00c00650-fd5e-4474-bbb1a27d636c36ef?video=00c00650-fd5e-4474-bbb1-a27d636c36ef

Stempfle J, Badke-Schaub P (2002) Thinking in design teams – an analysis of team communication. Des Stud 23(5):473–496

Sullivan L (1918) Kindergarten chats and other writings (Documents of modern art). Dover Publications, New York

Viollet-le-Duc EE (1874) English version: (trans: Towle G M) The story of a house. James R. Osgood and Company, Boston

Ward TB, Smith SM, Finke RA (1999) Creative cognition. In: Sternberg RJ (ed) Handbook of creativity. Cambridge University Press, Cambridge, MA, p 193

Chapter 14
Lesson Learned and Road Map

Roberta Ingaramo and Angioletta Voghera

Keywords Territorial design • Landscape design • Urban and architectural design

The waning of great transformation ideals based on independent architectural and territorial design gives way to a design capable of seconding the natural dimension, drawing on ecology to imagine living spaces in which nature regains its reference role in the construction of transformation scenarios. City as nature is the title of an issue of *Lotus International* in 2015: the city will possibly never be nature, but research into design which integrates various scales must be based on the reading and interpretation of the natural elements. This is even truer in the cases and reflections presented in this book, in which the relationship between the river and constructions takes the leading role.

Design is an effective tool for taking action in places where the crisis has hit hard, to grasp the potential of a territory and build transformation scenarios in parts that are capable, in the long term, of defining a future that is consistent with the identity of the territory and its resources.

In statement form, we have outlined several considerations which stem from theoretical reflections and the cases analysed, offering perspectives in terms of action and employment.

R. Ingaramo (✉)
Department of Architecture and Design, Politecnico di Torino,
Viale Mattioli 39, 10125 Turin, Italy
e-mail: roberta.ingaramo@polito.it

A. Voghera
Interuniversity Department of Regional and Urban Studies and Planning,
Politecnico di Torino, Viale Mattioli 39, 10125 Turin, Italy
e-mail: angioletta.voghera@polito.it

R. Ingaramo, A. Voghera (eds.), *Topics and Methods for Urban and Landscape Design*, Urban and Landscape Perspectives 19,
DOI 10.1007/978-3-319-51535-9_14

14.1 Trans-scalar Design Is Always Multidisciplinary

The definition of the boundaries of research for design is an integral part of the research itself, as is the outlining of the problems that have to be faced. Therefore, designs that concern extensive areas of land relate the work to the most varied scales, in a diachronic process that required different disciplines to deal with specific problems (integrated trans-scalar design, Ingaramo and Voghera Chapter; disciplinary bias, Plowright Chapter; urban heat island for sustainable planning, Mutani and Fiermonte Chapter; see also Mazzotta and Mutani 2015).

14.2 Design Has to Find the Right Scale in Order to Portray the Transformation and Guarantee Implementation

The characteristics of the problem form the reference to identify the operational scale for portraying the transformation. Perceiving the scale is the key to the success of a project. "In methodological terms, a permanent calibration is necessary" (Desvigne 2012, p.21) enforcing the need to deal with all the scales of design at the same time, but also obliging implementation of the actions in the long term, with actions that gradually become more and more detailed to allow the construction of parts of city and nature. A simultaneity of trans-scalar design requires adjustment of the eye to the scale of the landscape, when the operation is assessed looking at both the vast and local areas. This is a reflection that starts with architecture and continues to the plan, and it is a necessary condition for implementation (design in parts, Ingaramo Chapter; right "attitude" to work, Mellano Chapter; see also Mellano 2015; systemic actions and detailed designs, Voghera Chapter).

14.3 Architectural Design Builds Territorial Strategies

General strategies and single territorial actions require a reflection in terms of design which starts from the definition of every single architectural element of the city and nature.

Every element is an integral part of design scenarios which are part of a broader strategy, but which are implemented independently and have a more controllable spatial dimension. *Scenarios* offer spatial solutions that can be compared and discussed, which are incremental and open to alteration, through communication with the social actors, and are capable of adding tangibility to the aims and strategies of the plan (utopias and ecological planning, Palazzo and Diko Chapter; see also Palazzo 2008; complex, multi-scalar, multiplayer and collaborative design, Voghera Chapter).

14.4 Design Is an Evolutionary and Incremental Process

Design is an adaptive process which is implemented at different times, seconding the availability of financial resources and the changed conditions of transformation. Design does not define an imperative form of transformation but works on an optative basis, opening up to changes in time and space (flexible vision in the variable geometry and dimension framework of policies, Voghera Chapter; set of biases and a value system for design process, Plowright Chapter; see also Plowright 2014).

14.5 Design Is a Tool of Knowledge and Decision-Making

Design is a process which is heuristic, explorative, interpretative and transformative (evaluation as a knowledge process for gradual learning, Brunetta Chapter; argumentative approach, De Rossi, Armando, Giusiano Chapter).

14.6 Design Is a Tool for Sharing

Participation in design is a process of acquisition of knowledge and culture; of building relationships between the environment and its inhabitants, enabling the management of conflicts and the enhancement of the territory and landscape according to the hopes of the community; and of building empowerment.

Participation is an important part of the governance of "concerted" design, based on building consensus in a plural composition of conflicts and interests and of sharing aims, procedures and implementation processes. Involving public and private actors, it requires the necessary consideration of the relations between these actors as a precondition of the definition of the governance model (river contract – Berruti and Moccia Chapter; see also Moccia 2014; Voghera Chapter; river basin plan evaluation – Brunetta Chapter Brunetta 2015).

14.7 Design Creates Responsibility in the Actors

Implementation and management of design, components of its governance, are processes which change in time, redefining the contextual conditions, relations and strategies of the actors. The actors, an intrinsic component of design, play specific roles in implementation, sharing design choices, in a process of participation, right from the outset, and sometime having undertaken specific commitments (action contract – Ingaramo and Voghera Chapter 4; see also Ingaramo and Voghera 2012; river contract – Berruti and Moccia Chapter; see also Moccia 2014).

14.8 Design Redefines Boundaries and Institutional Architectures

Design and its spatial limits redesign institutional boundaries, either bottom up or top down, building territorial aggregations, opening up to synergies between territories and actors and creating the conditions for cooperation and competition (administrative complexity can be an advantage, successful from the bottom up – Sabaté Bel Chapter; see also Sabaté Bel and Warren 2014).

14.9 Design Defines Clear Morphologies and Structures

Design defines a singular and clear morphological design, with importance for the values that it builds, capable of combining specificities and diversities, expressing the physical structures of the landscape, infrastructure, relational spaces and constructions (limits, Ingaramo Chapter; tree-like landscape, De Rossi, Armando, Giusiano Chapter; see also De Rossi 2009; agro-urban cluster, Manzo Chapter; the boundary and sub-boundary of the heritage park, Sabaté Bel Chapter).

The design of “in-between” spaces is strategic, in that it is a place for the dialogic reconfiguration of the different elements and spatial systems (spaces in between – Ingaramo Chapter)

None of the considerations are absolute. They merely draw a framework in which research related to design attempts to build the physical transformation of the places where we live.

References

Brunetta G (ed) (2015) Smart evaluation in regional development. Territorial scenarios in Trentino, Italy. Ashgate, Aldershot

De Rossi A (ed) (2009) Grande scala. Architettura, politica, forma. List, Barcelona

Desvigne M (2012) The landscape as precondition. Lotus Int 150:20–27

Ingaramo R, Voghera A (2012) Planning and architecture. Searching for an approach. Alinea Intenational, Florence

Mellano P (2015) Per una identità dei luoghi. In: Canella G, Manganaro E (eds) Per una architettura realista. Maggioli, Sant’Arcangelo di Romagna, pp 166–181

Moccia FD (2014) Restauro paesaggistico ambientale. In: Petroncelli E (ed) Progetto paesaggio tra letteratura e scienza. Liguori, Naples

Mazzotta A, Mutani G (2015) Environmental high performance urban landcover: experimentations in Urban Barriera, 6th International Building Physics Conference, Turin 14–17th June 2015. Energy Procedia

Palazzo D (2008) Urban design. Mondadori Università, Milan

Plowright PD (2014) Revealing architectural design. Methods, frameworks and tools. Routledge, London

Sabaté J, Warren M (2014) Transforming Catalonia's industrial colonies. In: Albrecht H, Hansell F (eds) Industrial and mining landscapes within world heritage context. Sächsisches Industriemuseum. IWGT/TU Bergakademie, Freiberg, pp 103–115

Index

R. Ingaramo, A. Voghera (eds.), *Topics and Methods for Urban and Landscape Design*, Urban and Landscape Perspectives 19,
DOI 10.1007/978-3-319-51535-9

MIX
Papier aus verantwortungsvollen Quellen
Paper from responsible sources
FSC® C105338

If you have any concerns about our products,
you can contact us on
ProductSafety@springernature.com

In case Publisher is established outside the EU,
the EU authorized representative is:
Springer Nature Customer Service Center GmbH
Europaplatz 3, 69115 Heidelberg, Germany

Printed by Libri Plureos GmbH
in Hamburg, Germany